AF598097

ADVANCED MATERIAL AND DEVICE APPLICATIONS WITH GERMANIUM

Edited by **Sanghyun Lee**

Advanced Material and Device Applications with Germanium
http://dx.doi.org/10.5772/intechopen.73146
Edited by Sanghyun Lee

Contributors

Svetlana Kobeleva, Sergey Yurchuk, Iliya Anfimov, Aixa Gonzalez, Natalia Moreno, Patricia Cordoba Sola, Adrian Habanyama, Sanghyun Lee

First published in London, United Kingdom, 2018 by IntechOpen
IntechOpen is the global imprint of INTECHOPEN LIMITED, registered in England and Wales, registration number: 11086078, The Shard, 25th floor, 32 London Bridge Street
London, SE19SG – United Kingdom
Printed in Croatia

British Library Cataloguing-in-Publication Data
A catalogue record for this book is available from the British Library

Additional hard copies can be obtained from orders@intechopen.com

Advanced Material and Device Applications with Germanium, Edited by Sanghyun Lee
p. cm.
Print ISBN 978-1-78984-031-5
Online ISBN 978-1-78984-032-2

For EU product safety concerns:
IN TECH d.o.o., Prolaz Marije Krucifikse Kozulić 3, 51000 Rijeka, Croatia,
info@intechopen.com or visit our website at intechopen.com.

Meet the editor

Prof Sanghyun Lee is an assistant professor and CET program coordinator at Indiana State University. Before joining Indiana State University, he had more than 10 years of cutting-edge industry research experience such as in advanced CMOS technology, nonvolatile memory, and photovoltaics. He received his PhD degree in electrical and computer engineering from North Carolina State University. His research interests are emerging nonvolatile memory, post–silicon-era transistors, new types of solar cells, and energy harvesting devices.

Contents

Preface

The framework of this book was developed in the search for a new demand for materials that would facilitate our advanced technologies in the twenty-first century. Today, technology has become a ubiquitous part of our lives and consequently a critical component of all areas of our life. People wake up with checking up their emails on their smartphones and go to work in their car that has lots of sensors such as collision detection, parking assistant, lane departure warning, and so on. With the help of rapidly growing artificial intelligence technology, a self-driving autonomous car gives you the luxury to sit back in the car seat to get to the work. These sound like science fiction or movie scenes from a Hollywood blockbuster about the future but this is happening right now. Research and development of these technologies are moving at lightning pace, which is attributed to the semiconductor technology. After the birth of semiconductor technology in 1947, it was heavily dominated by silicon-based semiconductor technology for more than 60 years. However, the limitation of silicon technology invited a number of researchers to search for other alternatives including germanium. At the beginning of internet-of-things era (IoT), more autonomous systems were implemented in every corner of our lives and this strongly demands faster and lower power semiconductor device applications to meet our needs.

Germanium is an interesting material to explore in many applications. Although we rarely see them in our everyday life, they are embedded in many other applications where they play a critical role such as a main component of machines, tools, systems, instruments, and so on. The major applications for germanium are infrared optics, optical fibers, semiconductor devices, catalysts, and nuclear radiation detectors. At the beginning of a postsilicon era, most experts expect developing applications with new material alternatives; in particular, germanium beyond silicon technologies would eventually lead to a new technology revolution in the coming centuries.

In the fast-moving age of digital knowledge, it is almost impossible to follow up with all the new technologies. Furthermore, synchronized with these new technologies, a market keeps changing and transforming on a nearly yearly basis as we spend more time on our mobiles, tablets, laptops, and newly introduced gadgets. In the center of these changes, we are excited to present this book to readers in the belief that we can contribute to the process of deepening new knowledge.

Prof. Sanghyun Lee, PhD
Indiana State University
Terre Haute, Indiana, USA

Chapter 1

Introductory Chapter: Advanced Material and Device Applications with Germanium

Sanghyun Lee

Additional information is available at the end of the chapter

http://dx.doi.org/10.5772/intechopen.80872

1. Introduction

The search for new semiconductor materials began with new technology requirements in the early nineteenth century [1]. One of the pivotal discoveries was silicon (Si) and germanium (Ge), by Clemens Winkler in 1886. The current semiconductor technology is mostly based on Si material to fabricate integrated circuits (ICs) in the era of high-speed Internet-of-Things (IoT). Since Si transistors of ICs have faced physical limitations due to their fundamental properties, a number of researchers are actively searching for alternatives, which reignite the active study of Ge to break through the technology roadblock. The scope of this introduction is to describe the historical background of Ge materials from ores and their application to advanced devices such as photodetectors, solar cells, spintronics, IC, etc., which are essentially semiconductor applications in all areas of our current technologies [2]. Furthermore, this chapter will discuss the opportunities and challenges of Ge materials and advanced device applications for the next technology generation. The last part of this section will outline the topic of each chapter with some practical suggestions on how to efficiently utilize this book for readers.

2. Elemental semiconductor material

Nowadays, almost 95% of all the semiconductors are fabricated on Si material. Si semiconductors began to be used in the mid-1960s. The silicon devices demonstrate better and stable properties at room temperature. Furthermore, generating high-purity silicon wafer can be relatively easily achieved by the so-called Czochralski process. This is a method of crystal growth used to obtain single crystals of semiconductors, where high-quality silicon dioxide can be grown at room temperature [3–8]. From the economic perspective, high-purity Si for

device applications is much cheaper due to the fact that Si consists of 25% of the Earth's crust in the form of silica and silicates, which makes Si the second-abundant material after oxygen. As of now, Si technology is the leading technology surpassing all other material applications combined. However, a preferred choice of material for advanced ICs during the beginning of the semiconductor era in 1960 was interestingly Ge over Si after the invention of the first transistor with Ge in 1947. In the twenty-first century, the return of advanced Ge devices preparing post-Si device era invites us to look into advantages of Ge advanced devices, which makes it only possible with current state-of-art advanced technologies for at least next 50 years.

In 1947, two scientists at the Bell Telephone Laboratories, John Bardeen and Walter Brattain, made a triangular insulating wedge where two thin gold contacts with approximately 50-um-wide gap were glued on. They pressed this wedge into a slab of Ge and made a third contact on the bottom of the slab. They applied forward bias between one gold contact and the bottom of the Ge slab while applying reverse bias between the other gold contact and the bottom. This turned a small signal into a larger signal with the flow of current through this configuration, which had changed forever the history of semiconductors by inventing the first transistor—the amplifier and switch, arguably the most important invention of the twentieth century (see **Figure 1**).

At the critical juncture of the post-Si era, serious efforts searching for a new semiconductor material to replace long-standing Si devices began. In the past 10 years, most of leading semiconductor companies have begun to consider a certain change in components of their IC design such as the current-carrying channel, which is the very heart of a transistor. The idea is to replace the Si with a material that can move current at significantly greater rates. Compared to Si channels, alternative transistors with such channels could allow design engineers to design faster, denser, and low-power circuits, meaning better and cheaper smartphones, computers, and numerous IoT gadgets and applications in the market.

The existence and properties of such a material were first predicted by Dmitri Mendeleev in 1869, by filling a gap in the carbon family, in his periodic table of elements, located between

Figure 1. A stylized replica of the first transistor [2].

silicon and tin; therefore, it was called eka-silicon (Es), with estimated atomic weight of 72.0, which is not far off from 72.630, the standard atomic weight in modern chemistry [2, 9]. Although Ge is 50th in the relative abundance of elements in the Earth's crust, Ge came to be known relatively late in the history of chemistry due to the fact that it is rarely discovered in high concentration [2]. In 1886, a German chemist, Clemens Winkler was able to isolate it, and found it similar to antimony, named in honor of his home country.

Until the late 1930s, Ge was considered to be a poorly conducting metal [10], rather than a semiconductor, which made Ge economically insignificant. However, this had changed after World War II when Ge's semiconducting properties of diodes were found. In other words, the switching property of Ge diodes initiated the initial development of Ge devices [11, 12]. The first application was for the use in radar units as a frequency mixer element in microwave radar receivers by producing pure Ge crystal mixer diodes with the point-contact Schottky diode structure during the war period. During the post-war period, the development and manufacturing of solid-state Ge devices became a major stream in the semiconductor industry. From 1950 to the early 1970s, the Ge-related market increased for applications in transistors, diodes, and rectifiers [13]. For example, in the US, a few hundred pounds of production in 1946 greatly grew to more than 45 metric tons until 1960 in order to meet the market demand. However, soon after, high-purity silicon began replacing germanium in transistors, diodes, and rectifiers [13]. For example, the company that became Fairchild Semiconductor was founded in 1957 with the express purpose of producing silicon transistors. Silicon has superior electrical properties, but it requires much greater purity and that could not be commercially achieved in the early years of semiconductor electronics. Meanwhile, demand for germanium in fiber optics communication networks, infrared night vision systems, and polymerization catalysts increased dramatically. These end users represented 85% of worldwide germanium consumption in 2000 [12, 13].

Under the standard temperature (273.15 K) and pressure (105 Pa), Ge is a brittle, silvery-white, semi-metallic element [12]. As pure Ge is not mined as a primary material, Ge can be produced as a by-product of base metal refining [14]. Ge can be mostly found in the form of sphalerite zinc ores where it is concentrated in amounts as great as 0.3%, argyrodite (a sulfide of germanium and silver), and germanite (containing 8% of the element) [14, 15]. With sphalerite zinc ores, Ge concentrates are first purified using a chlorination and distillation process that produces germanium tetrachloride ($GeCl_4$) [15]. Germanium tetrachloride is hydrolyzed and dried, producing germanium dioxide (GeO_2), which is reduced with hydrogen to form Ge metal powder. Ge powder is cast into bars at high temperatures over 1720.85 F, which are treated by the zone-refining process to isolate and remove impurities. After this process, high-purity Ge metal bars are finally produced. Commercial Ge metal is often more than 99.999% pure. Zone-refined Ge can further be grown into crystals, which are sliced into thin pieces for use in semiconductors and optical lenses [15].

3. Applications and opportunities of germanium

In general, the United States Geological Survey (USGS) classified Ge applications into five groups such as IR optics (30%); fiber optics (20%); polyethylene terephthalate—PET (20%); electronic and solar (15%); and phosphors, metallurgy, and organic applications (5%) (see **Figure 2**).

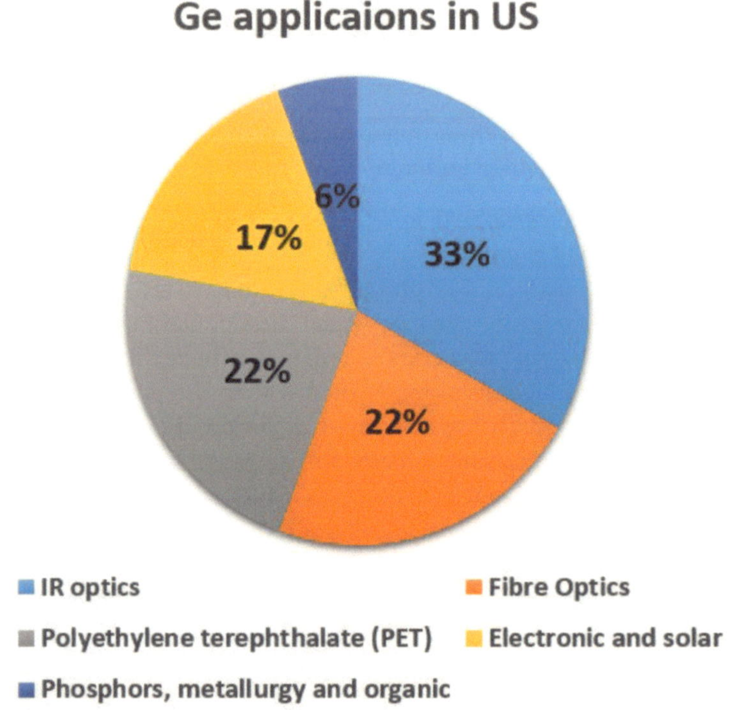

Figure 2. US Ge applications [12].

As mentioned earlier, zone-refined Ge crystals are grown and sliced to form lenses and window for IR and/or thermal imaging optical systems [15]. A major developer and customer is the military, for the application of advanced weapon systems such as small hand-held and weapon-mounted devices.

In fiber optics, telecommunication is possible by confining the light signal to their core, with Ge fibers acting as a waveguide for the electromagnetic light wave. Hence, the higher refractive index of the center of the fibers can improve the confinement of the light signal. Doping fused silica with Ge dopants can improve the refractive index in the silica glass of fiber optic lines by reducing signal loss (see **Figure 3**).

Regarding the production of PET plastics, roughly 17 metric tons of germanium dioxide is consumed each year as a polymerization catalyst. PET plastic is primarily used in beverage, liquid containers, and food.

In recent years, Ge has seen increasing use in precious metal alloys [12, 16]. In sterling silver alloys, for instance, it reduces firescale, increases tarnish resistance, and improves precipitation hardening. A tarnish-proof silver alloy trademarked Argentium contains 1.2% germanium [12, 16].

http://dx.doi.org/10.5772/intechopen.80872

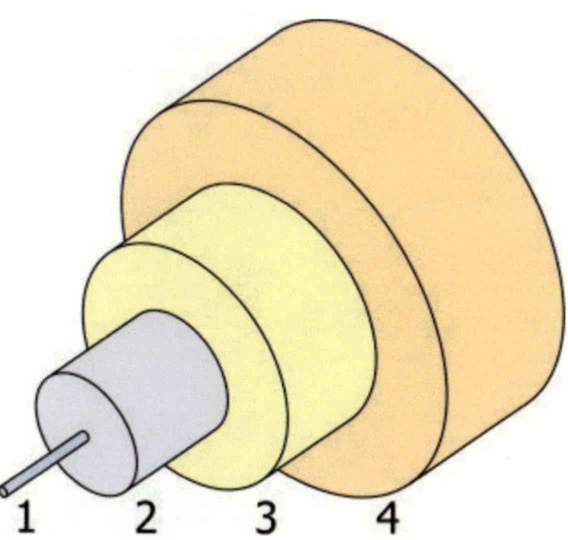

Figure 3. A typical single optical fiber, demonstrating core silica (1) with germanium oxide dopants, (2) cladding, (3) buffer, and (4) jacket [16].

As for spintronics, Ge is an emerging material for spin-based quantum computing applications. After finding the Ge property of spin transport at room temperature in 2010 [17], scientists recently showed very long coherence times of donor electron spins in Ge [16–18].

Soon after the birth of Ge transistors, Si transistors' replacement of Ge transistor happened in the early 1970s due to the reasons mentioned earlier. Today, however, scientists' efforts to achieve lower-power and higher-speed transistors have brought Ge back to the main interest of the semiconductor industry. By implementing Ge as current-carrying channel (channel) of the transistor, transistors are improved based on a fundamental property, which is the mobility of electrons and holes. Electrons move nearly three times as readily in Ge as they do in Si near room temperature. Furthermore, holes move about four times as readily in Ge. This is ultimately related to the difference in band structure between Ge and Si. Consequently, the faster these electrons and holes can move, the faster the resulting circuits can be. Since less voltage can be applied to draw those charge carriers along, circuits can also consume considerably less energy [19].

Since Ge band gap is small, 0.67 eV, Ge is transparent in the infrared wavelengths, which makes it possible to employ them in infrared spectroscopes and other extremely sensitive optical equipment such as infrared detectors. There are a number of infrared optical applications for Ge which can be readily cut and polished into windows and lenses [16, 20–22]. In particular, military applications rely on Ge optical properties. For instance, Ge is used in the front optic of thermal imaging cameras working in the 8–14-μm range for passive thermal imaging and for hot-spot detection in the military, mobile night vision, and firefighting applications (see **Figure 4**) [16, 20, 23, 24].

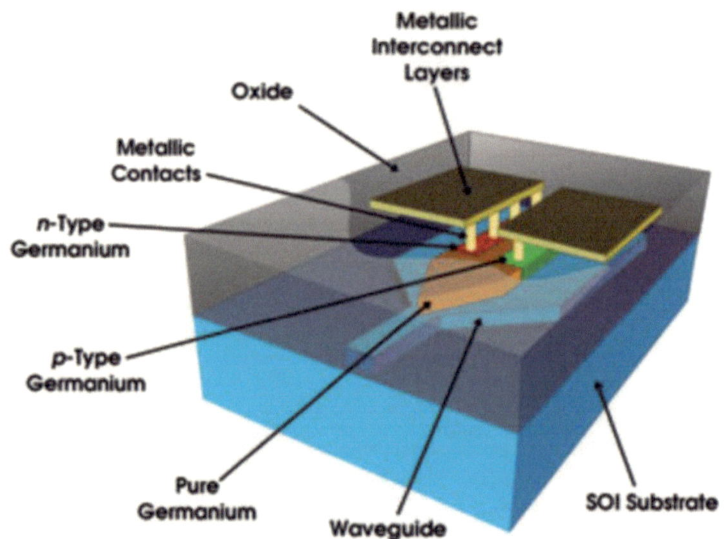

Figure 4. Germanium photodetector comprised of various layers of germanium [21].

4. Summary

We attempt to review germanium elemental materials and summarize the highlights of the history of germanium. Although the complete story of germanium would be lengthy, we have tried to highlight the material and device aspects of germanium. From the Ge application perspective, there are a number of examples; but the main applications could be categorized into five areas such as IR optics; fiber optics; polyethylene terephthalate—PET; electronic and solar; and phosphors, metallurgy, and organic applications. After the first invention of a transistor in 1947, which is arguably perceived as the most important invention in the twentieth century, and silicon replacement, germanium in electronics remarkably returns. The steady-fast increase of germanium application in IR and fiber optics is interesting as well. Furthermore, it is expected that more studies of germanium nanocrystals could contribute to the increased attention to research and development of new germanium applications in near future.

5. Outline of this book

As indicated in the title of this book, it will cover the detailed aspects of germanium. In particular, it is categorized into four sections, which describe critical aspects of germanium application in each field, including this first brief introduction of germanium. In the second part, this book will describe germanium material property and production from economical and engineering perspectives. In the third part, it will discuss germanium optics applications and diffusion characteristics during the process. In the final section with two chapters, this book will describe germanium microelectronic characteristics, in particular, interface characteristics impacted by the process. The contributing authors are experts in their field with great in-depth knowledge, which is contained in this book. The authors strongly feel that this contribution might be of interest to readers and help to expand the scope of their knowledge.

Author details

Sanghyun Lee

Address all correspondence to: sanghyun.lee@indstate.edu

Department of Electronics and Computer Engineering Technology, Indiana State University, Terre Haute, USA

References

[1] Smith RA. Semiconductors. 2nd ed. London: Cambridge University Press; 1979

[2] Wikipedia. History of the transistor [Internet]. Available from: https://en.wikipedia.org/wiki/History_of_the_transistor

[3] Haller E. Ge-based devices from materials to devices. Materials Science in Semiconductor Processing. 2006;**9**:408

[4] Teal GK, Sparks M, Buehler E. Single crystal germanium. Proceedings of the IRE. 1952;**40**:906-909

[5] Prabhakaran K, Ogino T. Oxidation of Ge(100) and Ge(111) surfaces: An UPS and XPS study. Surface Science. 1995;**325**:263-271

[6] Electronics History 4-Transistors [Internet]. [Updated: 2008]. National Academy of Engineering. Available from: http://www.greatachievements.org/?id=3967

[7] Greenwood NN, Earnshaw A. Chemistry of the Elements. 2nd ed. Oxford: Butterworth-Heinemann; 1997

[8] Holleman AF, Wiberg E, Wiberg N. Lehrbuch der Anorganischen Chemie. 102nd ed. Berlin/New York: de Gruyter; 2007

[9] Masanori Kaji DI. Mendeleev's concept of chemical elements and the principles of chemistry. Bulletin for the History of Chemistry. 2002;**27**(1):4-16

[10] Germanium for Electronic Devices [Internet]. W.K./The New York Times. May 10, 1953. Available from: https://www.nytimes.com/1953/05/10/archives/germanium-for-electronic-devices.html

[11] Computer History Museum. Semiconductor diode rectifiers serve in WW II [Internet]. [Accessed: Aug 22, 2008]

[12] USGS. U.S. Geological Survey. [Internet]. 2008. [Accessed: 2008]. Available from: https://minerals.usgs.gov/minerals/pubs/commodity/germanium/

[13] Emsley J. Nature's Building Blocks. 1st ed. Oxford: Oxford University Press; 2001

[14] Los Alamos National Laboratory. Periodic Table of Elements: LANL [Internet]. Available from: https://periodic.lanl.gov/32.shtml [Accessed: Aug 8, 2008]

[15] Terence Bell. Learn About Germanium [Internet]. [Updated: Aug 21 2017]. Available from: https://www.thebalance.com/metal-profile-germanium-2340135

[16] Wikipedia. Germanium [Internet]. [Updated: 2018]. Available from: https://en.wikipedia.org/wiki/Germanium

[17] Shen C, Trypiniotis T, Lee JY, Holmes SN, Mansell R, Husain M, et al. Spin transport in germanium at room temperature. Applied Physics Letters. 2010;**16**:162104

[18] Sigillito AJ, Jock RM, Tyryshkin AM, Beeman JW, Haller EE, Itoh KM. Electron spin coherence of shallow donors in natural and isotopically enriched Germanium. Physical Review Letters. 2015;**115**:24247601

[19] Peide DYe. Germanium Can Take Transistors Where Silicon Can't [Internet]. Available from: https://spectrum.ieee.org/semiconductors/materials/germanium-can-take-transistors-where-silicon-cant [Accessed: Nov 29, 2016]

[20] Moskalyk RR. Review of germanium processing worldwide. Minerals Engineering. 2004;**17**(3):393-402

[21] Photonics media. Closing in on Photonics Large-Scale Integration [Internet]. Available from: https://www.photonics.com/Articles/Closing_in_on_Photonics_Large-Scale_Integration/a31652 [Accessed: Dec 2007]

[22] Sclar N. Properties of doped silicon and germanium infrared detectors. Progress in Quantum Electronics. 1984;**9**:149-257

[23] Rogalski A. Infrared detectors: An overview. Infrared Physics & Technology. 2002;**43**: 187-210

[24] Otuonye U, Kim HW, Lu WD. Ge nanowire photodetector with high photoconductive gain epitaxially integrated on Si substrate. Applied Physics Letters. 2017;**110**:173104

Chapter 2

Germanium: Current and Novel Recovery Processes

Aixa González Ruiz, Patricia Córdoba Sola and
Natalia Moreno Palmerola

Additional information is available at the end of the chapter

http://dx.doi.org/10.5772/intechopen.77997

Abstract

Germanium (Ge) is considered a critical element due to its many industrial applications; Ge is a metalloid used in solar cells, fiber optics, metallurgy, chemotherapy, and polymerization catalysis. The main sources of Ge are sulfides ores of Zn, Pb, and Cu, coal deposits, as well as by-products and residues from the processing of these ores and coals (e.g., smelting flue dust and coal fly ashes). Indeed, over 30% of global Ge consumed come from recycling processes. The recovery of Ge from sulfide ores is mostly based on hydrometallurgical processes followed by a number of mass transfer techniques to concentrate Ge (e.g., solvent extraction). However, environmental-friendly extraction methods of Ge from coal fly ashes and copper smelting flue dust have recently been proposed in order to reduce environmental impacts. In addition, novel processes based on absorption of Ge with ribbon grass have become an interesting option not only to produce Ge but also to boost soil decontamination and biogas production. This chapter presents a general description of Ge occurrence, associations, and chemistry as well as a review of the current and novel recovery processes of Ge. The main sources of Ge and its main industrial applications are also discussed.

Keywords: germanium, residues, by-products, hydrometallurgical processes

1. Introduction

Germanium (Ge) was discovered in Freiberg in 1885. One year later, it was isolated from the uncommon mineral argyrodite (Ag_8GeS_6) by the chemist Clemens Winkler [1]. Ge is a chemical element with a grayish-white color whose position in the periodic table indicates that it has physicochemical properties similar to silicon (Si) and tin (Sn). Germanium has five

naturally occurring isotopes, $_{70}Ge$, $_{72}Ge$, $_{73}Ge$, $_{74}Ge$, and $_{76}Ge$, the latter being slightly radioactive, with a half-life of 1.58×10^{21} years [2]. However, $_{74}Ge$ is the most common isotope, having a natural abundance of approximately 36% [3].

Germanium is a scarce element in the Earth's crust (about 1.6 ppm Ge crustal average) that rarely forms its own minerals [6]. It often appears in the form of the oxide (GeO_2) or the sulfide (GeS_2) and in solution as germanic acid [7, 8]. However, most Ge is dispersed through silicate minerals due to the substitution of Ge^{4+} for the geochemically similar Si^{4+}. Ge is also associated with minerals or ores containing graphite (C), zinc (Zn), copper (Cu), iron (Fe), tin (Sn), and silver (Ag) [8]. Germanium can be classified either as a semimetal or metalloid because it shows both metal and nonmetal properties [4, 5]. As pure element, Ge has a metallic appearance at room temperature and behaves brittle with increasing mechanical deformation. In addition, Ge has a high refractive index and low chromatic dispersion and ability to form extended three-dimensional networks of Ge-O tetrahedra like Si-O [1, 9]. These physical properties determine the high economic importance of Ge and compounds in the industry sector. In this regard, the European Commission included Ge in a list of raw materials of critical concern for members of the European Union (EU) not only because of its high economic importance [10] but also because its industrial production is focalized in a small number of countries (mainly in China), and the world's demand pressure is increasingly growing.

2. Germanium-bearing minerals and residues

Nearly 30 minerals are known to contain Ge, mostly sulfides (**Table 1**). Germanium is a substituting element in Zn-sulfide structures (up to 3000 ppm in sphalerite and wurtzite ((Zn,Fe)S) and Cu sulfides (up to 5000 ppm in enargite, tennantite, bornite, and chalcopyrite) [3]. Germanium is also present as a substituting element in oxides (e.g., up to 7000 ppm in hematite (Fe_2O_3)), hydroxides (up to 5310 ppm in goethite (FeOOH)), phosphates, arsenates, vanadates, tungstates, and sulfates [3, 11].

Germanium can also occur in rare minerals such as argyrodite (Ag_8GeS_6), germanite ($Cu_{13}Fe_2Ge_2S_{16}$), renierite ($(Cu,Zn)_{11}(Ge,As)_2Fe_4S_{16}$), or briartite ($Cu_2(Fe,Zn)GeS_4$). Germanium-bearing ores are hosted in a variety of deposit that contains Au, Pb, and Ag, apart from those of Cu and Zn. Deposit types that contain significant amounts of Ge include volcanogenic-hosted massive sulfide (VMS), sedimentary exhalative (SEDEX), Mississippi Valley-type (MVT), Pb-Zn (including Irish-type Zn-Pb deposits), Kipushi-type Zn-Pb-Cu replacement bodies in carbonate rocks, polymetallic Zn-Sn vein, and coal deposits [9].

The VMS deposits are major sources of Zn, Cu, Pb, Ag, and Au and significant sources for Ge. They typically occur as lenses of polymetallic massive sulfides that form at or near the seafloor in submarine volcanic environments and are classified according to base metal content, Au content, or host-rock lithology [12]. There are close to 350 known VMS deposits in Canada and over 800 known worldwide. The most common feature among all types of VMS deposits is that they are formed in extensional tectonic settings, including both oceanic seafloor spreading and arc environments. Sedimentary-exhalative (SEDEX) deposits, on the other

http://dx.doi.org/10.5772/intechopen.77997

% germanium	Mineral name	Chemical formula	MW
69.41% Ge	Argutite	GeO_2	104.61
53.91% Ge	Eyselite	$Fe{+}{+}{+}Ge{+}{+}{+}{+}_3O_7(OH)$	401.40
45.27% Ge	Otjisumeite	$PbGe_4O_9$	641.63
35.78% Ge	Bartelkeite	$PbFe{+}{+}Ge_3O_8$	608.87
31.50% Ge	Stottite	$Fe{+}{+}Ge(OH)_6$	230.50
24.49% Ge	Carboirite-III	$Fe{+}{+}A_{12}GeO_5(OH)_2$	296.43
23.59% Ge	Krieselite	$(Al,Ga)_2(Ge,C)O_4(OH)_2$	230.81
22.36% Ge	Carboirite-VIII	$Fe{+}{+}(Al,Ge)_2O[(Ge,Si)O_4](OH)_2$	292.20
22.31% Ge	Brunogeierite	$(Ge{+}{+},Fe{+}{+})Fe{+}{+}{+}_2O_4$	244.11
18.57% Ge	Briartite	$Cu_2(Zn,Fe)GeS_4$	390.97
16.49% Ge	Barquillite	Cu_2CdGeS_4	440.38
13.42% Ge	Schaurteite	$Ca_3Ge{+}{+}{+}{+}(SO_4)_2(OH)_6 \bullet 3.(H_2O)$	541.06
10.83% Ge	Carraraite	$Ca_3Ge(OH)_6(SO_4)(CO_3) \bullet 12H_2O$	670.75
10.79% Ge	Maikainite	$Cu_{20}(Fe,Cu)_6Mo_2Ge_6S_{32}$	3296.63
10.15% Ge	Germanocolusite	$Cu_{13}V(Ge,As)_3S_{16}$	1609.66
10.03% Ge	Polkovicite	$(Fe,Pb)_3(Ge,Fe)_{1-x}S_4$	470.48
9.86% Ge	Ovamboite	$Cu_{20}(Fe,Cu,Zn)_6W_2Ge_6S_{32}$	3470.24
9.78% Ge	Morozeviczite	$(Pb,Fe)_3Ge_{1-x}S_4$	705.33
9.10% Ge	Germanite	$Cu_{26}Fe_4Ge_4S_{32}$	3192.14
7.89% Ge	Catamarcaite	Cu_6GeWS_8	902.31
7.76% Ge	Putzite	$(Cu_{4\cdot7}Ag_{3\cdot3})GeS_6$	925.86
7.62% Ge	Itoite	$Pb_3[GeO_2(OH)_2](SO4)_2$	952.35
7.21% Ge	Fleischerite	$Pb_3Ge(SO_4)_2(OH)_6 \bullet 3(H_2O)$	1006.40
6.58% Ge	Renierite	$(Cu,Zn)_{11}(Ge,As)_2Fe_4S_{16}$	1655.51
6.44% Ge	Argyrodite	Ag_8GeS6	1127.95
5.60% Ge	Calvertite	$Cu_5Ge_{0\cdot5}S_4$	495.06
2.90% Ge	Tsumgallite	$GaO(OH)$	100.05
2.71% Ge	Mathewrogersite	$Pb_7(Fe,Cu)Al_3GeSi_{12}O_{36} \bullet (OH,H_2O)_6$	2678.79
1.30%Ge	Colusite	$Cu_{12}\text{-}_{13}V(As,Sb,Sn,Ge)_3S_{16}$	1673.29
0.32%Ge	Cadmoindite	$CdIn_2S_4$	

MW: molecular weight.
Adapted from [13].

Table 1. Mineral species sorted by the element Ge.

hand, occur as tabular Zn-Pb-Ag deposits that contain laminated, stratiform mineralization and may be hosted in shale, carbonate, or carbonate- or organic-rich clastic rocks (siltstone and less commonly sandstone and conglomerate rich).

Most SEDEX deposits are hosted either by bimodal volcanic and clastic sedimentary sequence that is commonly metamorphosed to amphibolite-granulite facies, as at Broken Hill, Australia, or by basinal marine, fine-grained sedimentary rocks comprised mostly of carbonaceous chert, shales, and siltstones, less commonly by sandstones and conglomerates [14]. Mineralogy of the SEDEX deposits includes sulfides, carbonates, barite, and quartz. The most common sulfide mineral is pyrite, but the main ore minerals are invariably sphalerite and galena. SEDEX deposits account for 50% of the Pb and Zn reserves and about 25% of the global production of these metals [14]. There are more than 120 SEDEX deposits worldwide with known grade and tonnage figures, and of these 45 have geological resources greater than 20 million tons of Pb + Zn [15].

The Mississippi Valley-type (MVT) deposits can be found in the Gordonsville-Elmwood Zn-Pb district in Tennessee. These deposits, on average, have grades of 400 ppm Ge in Zn ore concentrate, while other MVT deposits in the USA may contain 50 ppm Ge in sphalerite. The Huize MVT deposit, which is located in China (Yunnan Province), is one of the largest MVT deposits in China and produces zinc-lead and Ag, Ge, and Cd by-products [16].

The most significant carbonate-hosted Zn-Pb-Cu deposits that contain notable amounts of Ge are the Kipushi deposit in the Democratic Republic of the Congo and the Kabwe deposits in Zambia [9]. Germanium averages 68 ppm in bulk samples in the Kipushi deposit and occurs substituted in sulfide minerals, although it sometimes occurs in separate Cu-Fe-Ge sulfide minerals [17].

Coal and lignite deposits are also a significant source of Ge. The Lincang lignite mine (Yunnan Province) produces 16 metric tons of high-grade GeO_2 annually, of which 90% is exported [11]. Germanium-rich coal seams are interblended with siliceous rocks that have oxygen and carbon isotope characteristics which suggested a hydrothermal origin. However, it has also been proposed that hydrothermal fluids were then discharged first as hot springs along fault zones into Miocene basins where the Ge was concentrated in lignite seams within stratiform siliceous and siliceous-limestone deposits [18].

Among the deposits containing well-constrained Ge reserves, sulfidic Pb-Zn (> 5000 tons) and high Ge lignite deposits (> 19,000 tons) constitute the two most important types of known Ge deposits [19].

Germanium can exhibit siderophile, lithophile, chalcophile, and/or organophile behavior depending on the geologic environment where it is hosted [11]. Thus, Ge shows a siderophilic behavior due to its relatively high Ge contents (up to 250 ppm) in Fe oxides such as Fe_2O_3 and Fe_3O_4 [8]. The lithophile behavior is shown by slight enrichment of Ge in the continental crust relative to the oceanic crust and the upper mantle, while the chalcophilic property of Ge is evident for its economic level in Zn- and Cu-rich sulfide hydrothermal systems. The organophile behavior of Ge, one of the highest affinities for organic matter of all the elements commonly associated with carbonaceous sediments, is marked from its enrichment in organic matter (coal and lignite deposits), which is comparable with some Zn-sulfide ores [2, 8, 11, 20, 21].

Because Ge occurs in coal deposits, by-products and residues from coal combustion and/or gasification should also be considered as Ge sources. Fly ashes (FAs) generated by combustion and gasification processes of certain coals may contain important amounts of Ge (**Table 2**). In the same way, Ge is retrieved as a by-product of sulfides ores (e.g., Zn and Cu-Zn-Pb ores); therefore, residues obtained from the ore processing, e.g., smelting flue dust, should also be considered as a potential source of Ge. By including these sources, the potential supply of Ge could exceed its current primary production [22].

Ge-bearing residues/by-products.	Ge mineral phases*	Ge content (mg/kg)	Reference
Zn refinery residue	n.d.	3620	[23]
Cu-cake	n.d.	700	[24]
coal fly ash	GeO_2	4986	[25]
Gasification fly ash	GeO_2, GeS_2, $GeSnS_3$	<500	[26]
Cu smelting flue dust	n.d.	417	[27]
Waste optical fibers	n.r	1100	[28]

*Mineral phases obtained through X-ray diffraction; n.d., not detected; n.r, not reported.

Table 2. Selection of residues and by-products with a high potential for Ge recovery.

3. Germanium production and its main applications

Germanium was initially used industrially in transistors due to its semiconductor properties. However, it was later replaced by silicon, which has better behavior with respect to temperature [9]. In 2016, Ge applications in solar cells, fiber optics, metallurgy, and chemotherapy and as catalyst for polymerization of polyethylene terephthalate (PET) comprised in 2016 almost 80% of the global consumption of Ge [20, 28–30].

Fiber optics are the major use of Ge worldwide since it is used as a doping element in optical fibers, which contain approximately 4% Ge, the rest being silicon oxide (SiO_2). Germanium increases the refractive index of the optical fiber which helps to contain the light within the fiber and enables the transmission of the digital signal. Germanium can also be used to make lenses and window panes for infrared detectors, infrared devices mainly destined to military guidance, and weapon-sighting applications and cameras because of its transparency to infrared radiation. It can, therefore, be used in numerous applications such as surveillance, night vision, and satellite systems [29–31]. With regard solar cell applications, Ge is used in high-performance multi-junction cells (typically III–V cells) in the domain of photovoltaics (PV) and in the bottom-cell part of triple junction PV, for the substrate, base, and emitter layers, because of its lattice constant, robustness, low cost, abundance, and ease of production.

Other diverse uses of Ge could be as an alloying element (0.35%) for Sn, or Al-Mg alloys, to increase their hardness; soldering material (12%Ge/88%Au) for gold-based dental prosthesis; luminescent material; photographic and wide-angle lenses; ceramics, with Na_2O/TiO_2 or K_2O/Ta_2O_5; gamma-ray detector $Bi_2(GeO_3)_3$; bismuth germanate oxide crystals (BGO- $Bi_4Ge_3O_{12}$) for various detection technologies (scintillation, tomography, gamma spectroscopy); fluorescent paint ($MgGeO_3$); superconductors (Nb_3Ge); thermocouple; and thermoelectricity. Germanium dioxide is also used as a polymerization catalyst in the production of PET, giving rise to a wide range of PET bottles and containers [29–31].

As explained in last sections, Ge recovery is associated with currently produced Zn and Cu-polymetallic ores and coal deposits [11, 31–33]. None of the Ge-bearing minerals is mined solely for its content, and most of the recovered Ge is a by-product from ores and coal processing [1]. Therefore, the extraction of Ge is mostly carried out through the typical extraction methods by mining facilities (pyro- and hydrometallurgy) [1, 25]. **Figure 1** shows the pathway processing for Ge recovering either by Zn refining residues and scrap. In general, after physical separation or pyro- and hydrometallurgical processes, a concentrate of Ge with around 30% content is obtained. Thus, the Ge concentrate, regardless of its source, is chlorinated, distilled, and purified to form the first usable product, $GeCl_4$, which is primarily used in fiber-optic cable production [3]. Germanium tetrachloride can be hydrolyzed and dried to

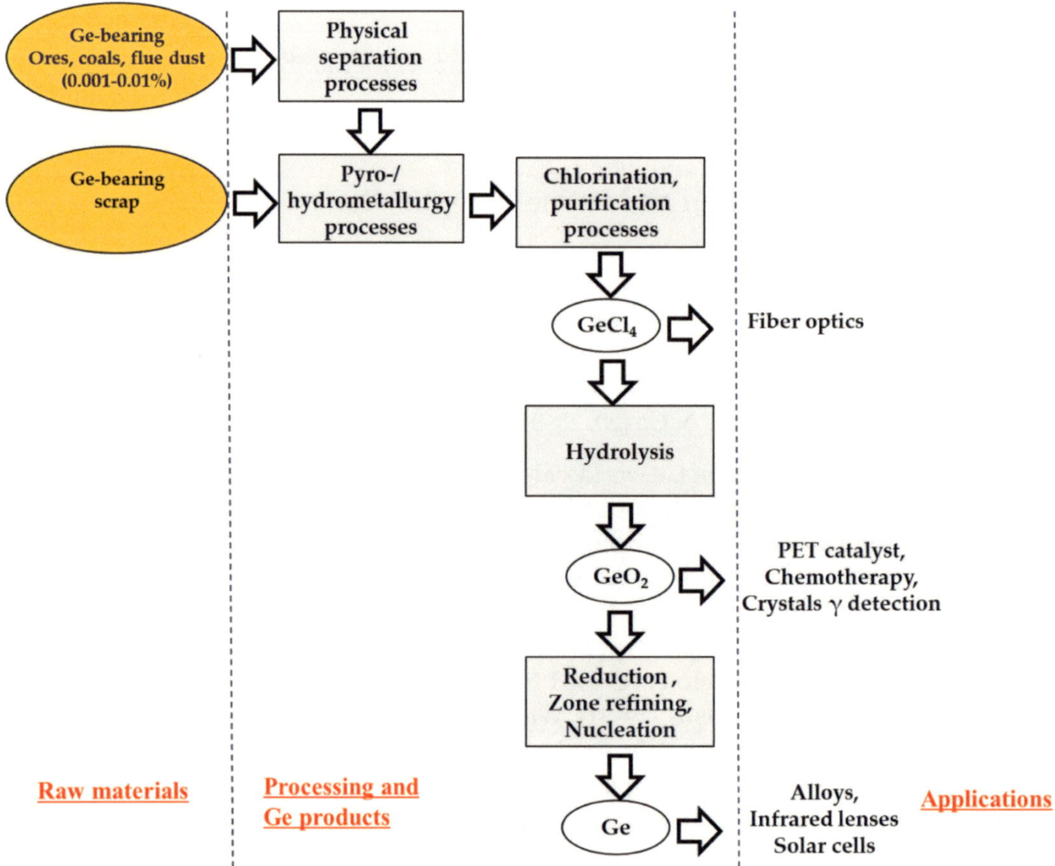

Figure 1. Germanium processing pathway, modified from Melcher and Buchholz [3].

produce GeO_2, which is used in the manufacture of certain types of optical lenses and as a catalyst in the production of PET resin. Germanium metal powder is produced through the reduction of GeO_2 with hydrogen.

In 2017, the worldwide production of Ge was estimated to be about 134,000 kg that is mainly recovered from Zn concentrates, coal deposits, coal fly ashes, and recycled materials [20]. While several authors have reported an increase (~30%) of the Ge production in the last decade, it is known that Ge reserve is scarce and it is estimated to be 8600 tons [9, 28, 34, 35]. The contradiction between the increasing consumption and the scarce reserve of Ge is becoming more notorious and has been contributed to a strong recycling process for Ge. In 2016, about 30% of the total Ge consumed was supplied from scrap (recycled materials), e.g., from windows in decommissioned tanks and military vehicles. In special, recycling rates for fiber-optic scrap are reported as high as 80%. As a consequence, about 50% of the Ge metal used for electronic and optic are recycled in short cycle [36, 37].

As shown in **Figure 2**, the worldwide production of Ge is led by China (65.7%) followed by Russia (5%) and other countries such as Canada, Belgium, and Germany (30%) [20]. In China, Ge use in fiber optics increased substantially from 2012 to 2016 which supposed the highest consumption growth of Ge. Moreover, countries such as the USA and China treat Ge a strategic reserve, due to important value for the high-tech industry for civilian and military purposes [38]. Several authors indicate that reliable information about global Ge prices for public domain is very little published [1, 38, 39]. From the point of view of the authors, the last statements have partially contributed to the extremely global Ge price fluctuations, exceeding in last year 1300 US $/kg for Ge metal only in the USA [20].

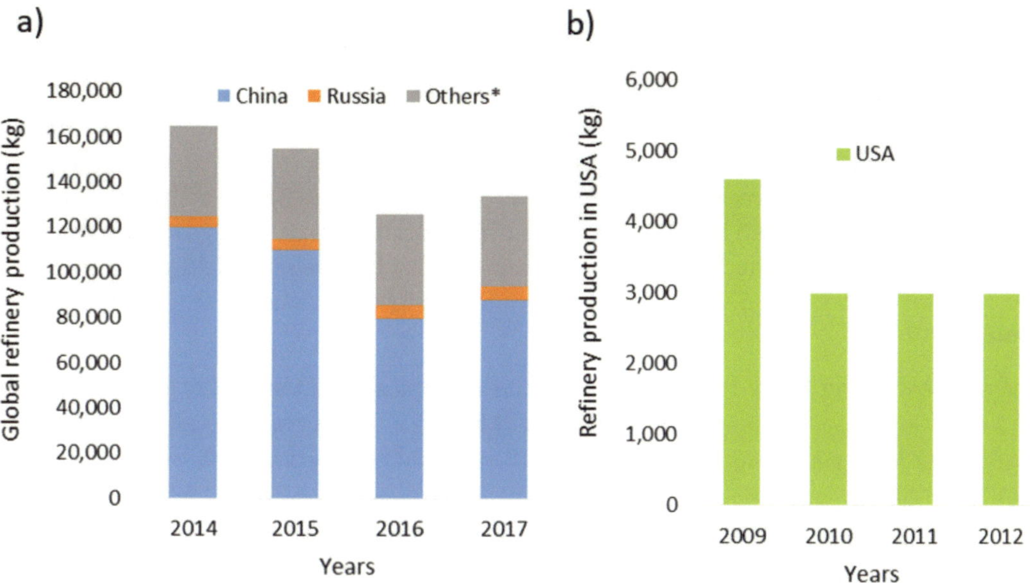

Figure 2. Germanium refinery production of (a) global main producers between 2014 and 2017 excluding the USA—* includes Belgium, Canada, Germany and others [20, 29, 30, 34]—And (b) of USA between 2009 and 2012.

4. Current and novel recovery processes for germanium

Several processes have been recommended for Ge recovering based on its content, chemical species, and mineral phases that are present in by-products from Zn, Cu, and Pb ores processing and from coal combustion and gasification. In general, hydrometallurgical processes are favored, because almost all Ge extracted is mainly concentrated by processes based on mass transfer operations. Nevertheless, it is difficult to find extractants which meet the following characteristics: (i) selective, (ii) cost-effective, (iii) eco-friendly, and (iv) commercially available. In the next sections, the authors aboard main extraction methods used for the recovery of Ge from different sources and problems associated.

4.1. Recovery of germanium from Zn ore processing

Nowadays, Zn ore processing is the main source of Ge as Zn ores have large and recoverable quantities of Ge. Zinc refinery residues, which are the typical by-products of hydrometallurgical zinc processes, usually contain between 0.2–0.5 wt% Ge and 0.3–0.4 wt% Ga with Zn, SiO_2, Cu, Fe, and Pb as the main components [39, 40]. However, on a global scale, as little as 3% of the Ge contained in Zn concentrates is recovered since it can also have a negative impact on Zn recovery, detracting from the core business for refineries [20, 41]. As a consequence, except the Chinese refineries, only two Zn refineries currently extract Ge as part of their operations [36].

Several studies have been conducted on the behavior of Ge for the effectiveness of its recovering from by-products and residues of Zn processing [23, 42–45]. In 1987, a reductive SO_2 leaching process as an alternative for Ga and Ge recovery from Zn leaching residue was investigated [42]. Only 57% of Ge was extracted, which was mainly attributed to the formation of silica-germanium gel; H_4GeO_4 and H_4SiO_4 were shown to hydrolyze the mixed polymers [46–48]. A higher yield for Ge was observed with an alkaline process used to treat Zn refinery residues. However, the authors also detected that Si, Pb, and Al hinder the recovery and purification of Ge [44, 49]. Recovery of Ge from zinc refinery residues has more frequently been carried out by leaching with H_2SO_4 being the resulting solution treated with solvent extraction (SX). Currently, synergistic SX (SSX) has been proved to increase the yield of Ge recovery and purification. Some of SX and SSX processes for Ge recovery studied are summarized in **Table 3** [50–58].

Chelating extractant Kelex 100 was primarily used for the separation of Ge^{4+} from Zn, reaching Ge extraction 98% in a solution of 156 g/L H_2SO_4 [50]. However, slow-phase disengagement and a high concentration of NaOH were required to strip the Ge^{4+} from the loaded organic solution.

Moreover, two studies achieved a good separation of Ge from Cu, Ni, As, Cl, and Fe^{2+} with a H_2SO_4 solution of 100 g/L by the use of LIX 63 [51, 52]. In 1984, a SX system, consisting of LIX 63 and LIX 26, was used for the extraction of Ge from a solution containing Ge^{4+} (3.5 g/L), arsenite (0.8 g/L), and Fe^{3+} (1.5 g/L) [53]. Over 99% of Ge^{4+} was extracted with 99% (v/v) LIX 63 and 1% (v/v) LIX 26 in four stages at lower acidity (50 g/L H_2SO_4) compared with using LIX 63 alone (>90 g/L H_2SO_4).

Extractant	Main issues of process	Reference
	Solvent extraction (SX)	
Kelex 100	• Good separation of Ge from Zn, Cd, Ni, Co, and As • Poor phase separation in stripping • High concentration of NaOH required for stripping	[50]
LIX 63	• Good separation of Ge from Zn, Cu, Ni, As, Fe(II), and Cl • Low extraction efficiency • Slow extraction kinetics	[52, 53]
H106	• Ga and Ge co-extraction • Selective stripping • H106 is not commercially available	[55]
G315	• 95% Ge extraction efficiency at a low acidity • G315 is not commercially available	[56]
	Synergistic solvent extraction (SSX)	
D2EHPA + TBP	• TBP improves extraction efficiency and phase separation • High concentration of NaOH for stripping	[57]
LIX 63 + LIX 26	• Increased Ge extraction by addition of LIX 26	[53]
LIX 63 + O.P	• Good selective Ge extraction and high efficiency • Fast degradation of LIX 63 by the acidity of O.P acid	[58]

O.P: organophosphoric acid.
Adapted from [43].

Table 3. Summary of investigated SX or SSX systems for Ge recovery.

Other SX systems consisting of LIX 63 and D2EHPA, M2EHPA, and OPAP were also developed to recover Ge^{4+} from H_2SO_4 solutions (75 g/L) containing Zn^{2+}, arsenate, Cd^{2+}, Sb (V), In^{3+}, Cu^{2+}, and Fe^{2+} using a multistage counter-current process. In this case, over 95% of Ge^{4+} was extracted using a mixture of 25% LIX 63 and 75% D2EHPA [52].

A technology for In^{3+}, Ge^{4+}, and Ga^{3+} recovery from a H_2SO_4 leach solutions of Zn residue was also proposed [54]. Indium was first separated from the solution using SX with 30% D2EHPA in kerosene, while Ge (97%) and Ga (95%) were co-extracted with the SSX system consisting of 20% D2EHPA and 1% YW100 in kerosene at low pH values. However, the authors found that YW100 is not commercially available and the process is not eco-friendly.

The D2EHPA extractant was also developed for selective extraction of In^{3+} and Fe^{3+} and used H106 for co-extraction of Ga^{3+} and Ge^{4+} from a solution from H_2SO_4 leach solutions of a cementation residue in a mini pilot plant scale. Recoveries of 91%, 94%, and 93% for In, Ga, and Ge, respectively, were achieved [55].

The extractant G315 was tested to recover Ga and Ge from solutions containing Zn^{2+} (22.7 g/L), Ge^{4+} (0.1 g/L), Ga^{3+} (0.3 g/L), and Fe^{3+} (2.2 g/L), in 40 g/L H_2SO_4 at an aqueous/organic (A/O) phase ratio of 1:2. The extraction of Ge achieved 94.6%, but the structure or type of the extractant was not disclosed [56].

For the separation of Ge^{4+} from Zn^{2+}, Ga^{3+}, and Fe^{3+} in solutions with a high acidity (80 g/L H_2SO_4), a SSX system was also used [57]. The process consisted of 30% (v/v) D2EHPA and 15% (v/v) TBP. The extraction efficiency was 94.3% in two stages, and the strip efficiency was almost 100% using 250 g/L NaOH at an A/O phase ratio of 1:2.

A single contact system consisting of 10% LIX 63 and 2% Ionquest 801 to recover Ge from a synthetic leach solution of Zn refinery cementation residues was used [43]. Over 68% Ge^{4+} was extracted at a low pH at an A/O ratio of 1:1 and 40°C. Almost 73% Ge^{4+} was stripped with 0.5 M NaOH and 1.0 M Na_2SO_4.

A common disadvantage of the above SX or SSX processes for Ge^{4+} extraction is the use, for instance, of strong NaOH for stripping and of some reagents that are not commercially available. Therefore, more efficient and effective SX or SSX systems for the recovery of Ge are required using commercially available reagents.

4.2. Germanium from coal combustion and gasification fly ashes

Coal plays an essential role in our global energy scheme for power generation. There are 1,139,331 million tons of proven coal reserves worldwide, sufficient to meet 153 years of global production which makes coal a reliable source [59, 60]. However, coal is currently a target to accomplish with the Paris climate agreement for both countries and companies which has caused a decline in coal production and consumption. Pulverized coal combustion (PCC) is the most widely used technology for coal power generation and, in a lesser extent, integrated gasification combined cycle (IGCC). In both processes, coal with a proximate Ge content <100 ppm can either vaporize totally and then be easily adsorbed on the finest coal fly ash (FA) particles during flue gas cooling or vaporize partially and enrich in both the coal FAs and, in a lesser extent, bottom ashes or slags (**Figure 3**). FAs are normally captured in particulate control devices with a high efficiency (>90%), but a small fraction of them may reach the flue-gas desulfurization (FGD).

Although part of the FA components may dissolve in the aqueous phase of the sorbent slurry when the flue gas passes through the sprayers, remaining in the FGD, the content of Ge in the FGD by-products (water effluent and FGD-gypsum) is not significant [60, 61]. Coal FAs are regarded as the main output stream of Ge.

The current annual production of coal FA worldwide is estimated to be around 750 million tons, and this is anticipated to increase in the near future [62]. The average content of Ge in coal FA is approximately 18 mg/kg, but as some research showed, it can reach 420 mg/kg [26, 63]. Therefore, one attractive source of Ge comes from coal FAs [64–66]. Coal combustion FA is a fine powder made up of spherical high Si-Al-Ca-K-Fe-Ti-Mg vitreous particles with Fe oxides and Al-Si species and irregular unburned coal and ash particles (**Figure 4**). It is generally accepted that vitreous FA particles consist of a relatively pure Al-Si-Ca-K-Fe glass within on which mullite crystals form a network [67, 68]. Furthermore, other types of particles such as calcite ($CaCO_3$), lime (CaO), quartz (SiO_2), and gypsum ($CaSO_4 \cdot 2H_2O$) are formed.

During coal gasification, most of the mineral matter of the coal is transformed and melted into slag. As opposed to PCC, coal gasification produce very little FA (10–15%), which is

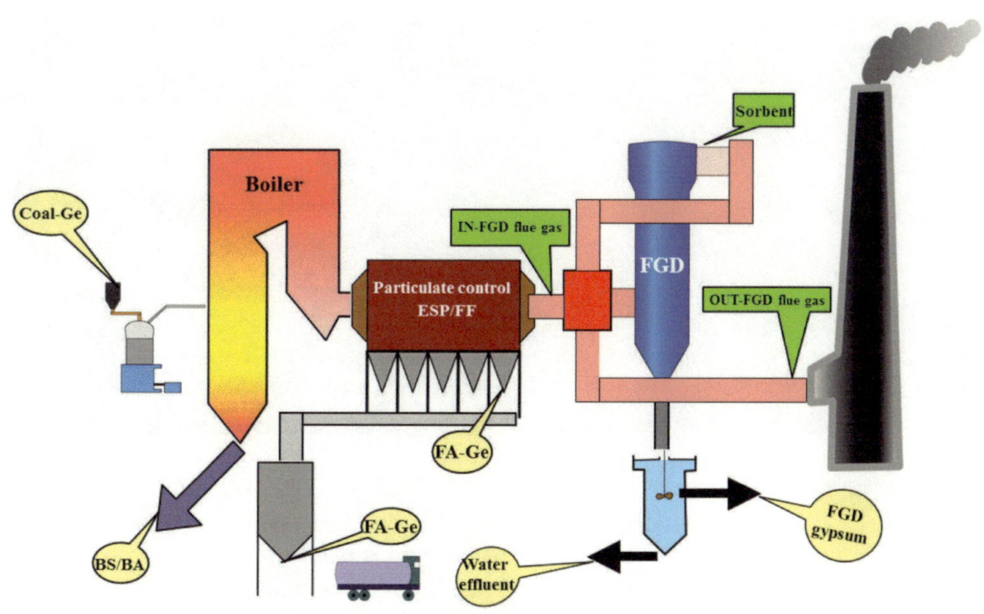

Figure 3. Configuration of a PCC power plant and partitioning of Ge.

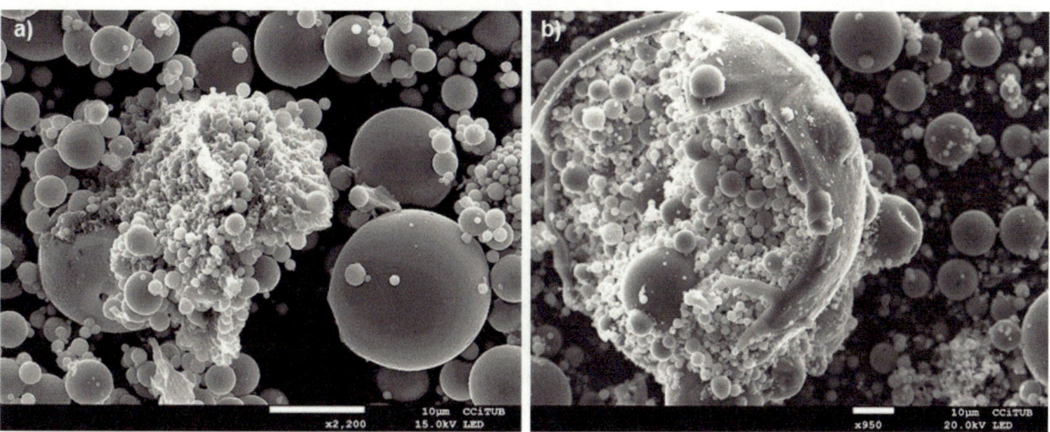

Figure 4. Scanning electron microscope photomicrographs of (a) FAs from coal and petroleum coke combustion and (b) FAs from PCC [61].

characterized by a predominant alumina-silicate glass matrix and a wide variety of crystalline-reduced species, mainly sulfides, because of the low levels of O_2 during coal gasification. **Figure 5** shows the typical round morphology of IGCC FA particles [26].

Germanium production from coal FAs usually consists of two stages. The first step creates a concentrate and the second is the actual recovery. The first published studies on Ge recovery from coal FAs were those based on pyro-metallurgical practices [70] but at the present moment are not applied due to the high economic and environmental cost [1].

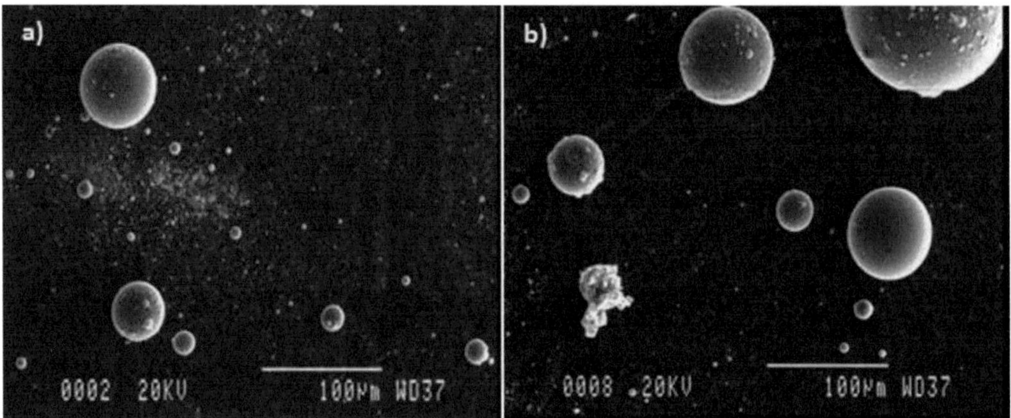

Figure 5. Scanning electron microscope photomicrographs of IGCC FAs [69].

In 1998, a hydrometallurgical procedure for extracting Ge from coal FAs based on the leaching of FA with H_2SO_4 and NaOH followed by ion flotation separation of Ge was prompted. A recovery of 100 g of Ge per ton of coal FA was achieved between 70 and 200°C [71]. Other studies implemented the ion flotation method using a mixed HCl/HNO_3 solution and cetylpyridiniumchloride as a surfactant [72]. The authors obtained 95–100% yield of Ge after 3–5 min of flotation. A research about tannin (poly-hydroxy polyphenols) precipitation to extract Ge was proposed in 2008 [73]. The tannins were capable of forming chelates with Ge ions, resulting in low grade of precipitated tannin-germanium complex. Currently, research is aimed at developing methods for increasing FA utilization focused on reducing the concentrations of heavy metals and at the same time obtaining higher added value products.

The occurrence of Ge as water-soluble species such as GeS_2, GeS, and hexagonal-GeO_2, in the FAs generated in an IGCC power plant, prompted the study of Ge recovery from coal FAs using pure water in an attempt to develop an extraction process of a low cost and environmentally able to culminate in a commercial Ge end-product [26]. Results revealed high recovery efficiency (up to 86%) at 90°C, indicating that the extraction temperature was the most important parameter in the process. These results led authors to conduct research toward the study of enrichment and precipitation methods for Ge recovery such as ion flotation, adsorption on activated carbon, and/or SX. Adsorption and SX were the methods that allowed achieving Ge-bearing solutions with 256 and 1623 mg/L, respectively [74, 75].

In 2006, a study evaluated the selectivity of the process developed by [26] for the recovery and purity of Ge by ion flotation tests on the leachates arising from the water extraction of Ge, using pyrogallol, catechol, hydroquinone, and resorb in complexing agents at a pH range of 4–7. Pyrogallol or catechol as complexing agents and dodecylamine as a surfactant showed the isolation of the Ge complex recovering 100% of Ge in 30 min [76].

A complex water leaching process consisting of Ge complexation with catechol followed by SX system, reaching an extraction yield of 95%, was published [77]. In 2014, same authors optimized Ge and Ga recovery from coal FA using different extractants in a wide range of extraction conditions. High extraction yields of Ge (up to 90%) and Ga (up to 82%) were obtained using weak oxalic acid and sulfuric acid solution, respectively, within 1–2 h extraction period [78].

Recently, the recovery of Ge from coal FA using vacuum reduction metallurgical process was investigated [38]. Its principle is that the saturation vapor pressure of metal under the vacuum condition is lower than normal pressure to separate metals. These authors achieved a 94.6% recovery of Ge from the coal FA at a temperature of 900°C, 10 Pa, and heating time of 40 min. These studies confirm that coal FA can be explored as an alternative source of Ge and other valuable elements and minerals. The recovery of Ge from coal FA is suitable, reducing disposal costs of coal FA. However, further research is needed to develop adequate results in terms of selective recovery and purification.

4.3. Germanium recovery from copper smelting flue dust

Flue dust from Cu smelting has also been suggested as a potential source of Ge, since relatively high contents of this element may be present in Cu-sulfide ores [27]. Copper is extracted from the ore through hydrometallurgical and/or pyrometallurgical processes. The selection of the process is determined by the Cu minerals bound to ores, being Cu sulfides predominantly treated by pyrometallurgical process and Cu oxides by hydrometallurgical process. However, the pyrometallurgical process is the most commonly employed technology for Cu [79].

The processing of Cu minerals, associated to sulfides by high temperatures, produces several residues [80]. Among the residues, dust generated from physical process, flue dust and slags from smelting process, and sludge from electrowinning process are those with potential for the Ge recovering.

Although the qualitative and quantitative phase determination of dust from Cu smelting depends on the compositional characteristics of the fed into smelting furnaces, temperature and oxidative conditions inside the furnace, and equipment, the recovery of rare and precious metals such as Ge from flue dusts has not been a widely studied subject in the literature. Most of the research on flue dust composition from Cu smelting have been focused on As, Zn, and Pb since the main Cu-ores present elevated amounts of As, Zn, Cd, and Pb which are potentially hazardous to human health or the environment [27].

Font et al. (2011) evaluated for the first time the potential of Cu smelting flue dust (Cu-SFD) as a source of Ge and the possibilities to apply extraction and recovery methods similar to those patented for coal gasification FA [81]. These authors reported Ge concentrations ranging from 417 to 1375 ppm in flue dust samples with Ge extraction yields from 73 to 99%. In 2017, Chilean Cu-SFD was characterized and evaluated for the potential extraction of Li, Rb, and Ge with different chemical leaching agents [27]. The authors found high extraction yields for Ge, Li, and Rb using pure water as extractant at 25°C. Both studies suggest that Ge may occur in the form of highly soluble minerals and that Cu-SFD can be regarded as a promising source of elements with high added value such as Ge.

4.4. PhytoGerm

Phytomining is an extraction process in which metallic substances in soils or sediments are absorbed by plants [82]. With this in mind, PhytoGerm project emerged in the framework of the r³-initiative for tech metals and resource efficiency subsidy program of the German Federal Ministry of Education and Research whose goal was to find a plant species that concentrates Ge in aerial plant biomass, which grows well on poor soils and contaminated industrial sites.

The ribbon grass was selected as suitable for the PhytoGerm project. Ribbon grasses grow well on prolific siliceous soil, and due to the similar chemical properties of Ge and Si, the plant can also absorb Ge [82]. The concept of PhytoGerm project was to make use of elevated Ge levels of tailings from Zn mining sites, thus allowing the plants to accumulate sufficiently high amounts of Ge in order to achieve high yields during the extraction process [83]. The case study developed by the authors assumed that 13,636 tons/year of ribbon grass would be obtained from several cultivation areas, which is the amount needed to utilize an average 500 kW biogas plant. Along the process diagram showed in **Figure 6**, 4112 tons/year of biomass are available for Ge extraction. Once Ge is accumulated in ribbon grass plants, the solid biomass is at first dried and thermally processed in a biomass power plant. The residuals of the combustion process are ashes and FAs, enriched with Ge, with an annual output of approximately 280 tons. The investigated process route ends with producing 3.9 kg of powdery GeO_2 per year.

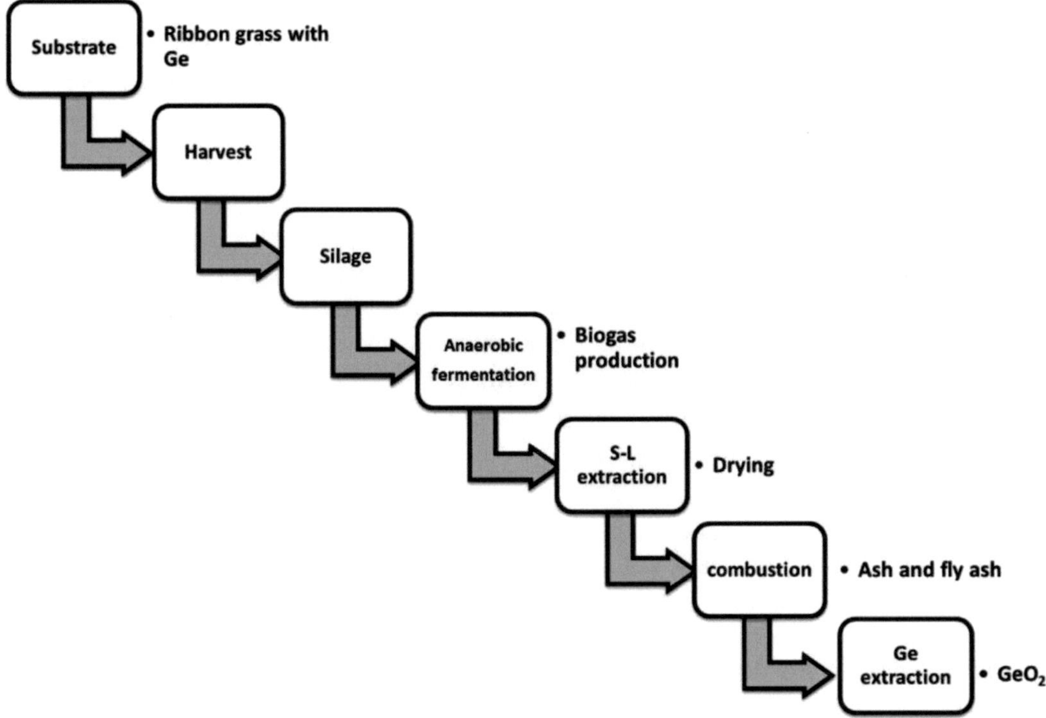

Figure 6. PhytoGerm process diagram. Adapted from [82].

5. Conclusion(s)

Nowadays, germanium is considered a critical element and also a strategic reserve for high-tech industrial applications in several countries. Germanium is used in solar cells, fiber optics, metallurgy, chemotherapy, and polymerization catalysis. Mainly sources of Ge are associated to sulfide ores (e.g., Zn, Pb, and Cu), coal deposits, and also residues from the processing of these ores and coals. Indeed, about one third of global germanium produced come from recycling processes. While the recovery of Ge from sulfide ores presents disadvantages related to the hazardous nature of organic extractants and high acidity of extractant solutions, the occurrence of Ge as water-soluble chemical species in coal gasification and copper smelting fly ashes allows the application of novel extraction methods with water at temperatures <100°C. This approach appears to be a feasible recovery and less harmful novel extraction method for environment, which suggests that both residues are promising sources for Ge. PhytoGerm which is based on absorption of Ge with ribbon grass on soils contaminated with Zn refinery residues results in an energy-efficient and eco-friendly recovery process for Ge.

Acknowledgements

The authors thank the support of FONDECYT under grant 11150088. The authors also gratefully acknowledge the Institute of Environmental Assessment and Water Research, Spanish National Research Council (IDÆA-CSIC).

Author details

Aixa González Ruiz[1*], Patricia Córdoba Sola[2] and Natalia Moreno Palmerola[2]

*Address all correspondence to: agonzalez@uct.cl

1 Department of Industrial Processes, Catholic University of Temuco, Temuco, Chile

2 Geosciences Department, IDAEA-CSIC, Barcelona, España

References

[1] Moskalyk RR. Review of germanium processing worldwide. Minerals Engineering. 2004;**17**:394-402. DOI: 10.1016/j.mineng.2003.11.014

[2] Meija J, Coplen TB, Berglund M, Brand WA, De BP, Gröning M, Holden NE, Irrgeher J, Loss RD, Walczyk T, Prohaska T. Isotopic compositions of the elements 2013 (IUPAC technical report). Pure and Applied Chemistry. 2016;**88**:293-306. DOI: 10.1016/j.chemgeo.2017.03.013

[3] Melcher F, Buchhloz P. Germanium. In: Gunn GA, editor. Critical Metals Handbook. John Wiley & Sons; 2014. pp. 177-203. DOI: 10.1002/9781118755341.ch8

[4] Germanium. Open chemistry data base [Internet]. Available from: https://pubchem.ncbi.nlm.nih.gov/compound/germanium#section [Accessed: 12-01-2018]

[5] Silicon. Open chemistry data base [Internet]. Available from: https://pubchem.ncbi.nlm.nih.gov/compound/silicon#section [Accessed: 12-01-2018]

[6] Bernstein LR. Germanium geochemistry and mineralogy. Geochimica et Cosmochimica Acta. 1985;**49**:2409-2422. DOI: 10.1016/0016-7037(85)90241-8

[7] Rosenberg E. Germanium: Environmental occurrence, importance and speciation. Reviews in Environmental Science and Biotechnology. 2009;**8**:29-57. DOI: 10.1007/s11157-008-9143-x

[8] Meng YM, Hu RZ. Minireview: Advances in germanium isotope analysis by multiple collector-inductively coupled plasma-mass spectrometry. Analytical Letters. 2018;**51**:627-647. DOI: 10.1080/00032719.2017.1350965

[9] Shanks PWC III, Kimball BE, Tolcin AC, Guberman DE. Germanium and indium. In: Schulz KJ, DeYoung JH Jr, Seal RR II, Bradley DC, editors. Critical Mineral Resources of the United States—Economic and Environmental Geology and Prospects for Future Supply. Virginia, USA: USGS; 2017. pp. 1-27. DOI: 10.3133/pp1802I

[10] European Commission. Critical raw materials for the EU [Internet]. Available from: http://ec.europa.eu/enterprise/policies/rawmaterials/documents/index_ en.htm [Accessed: 20-03-2018]

[11] Höll R, Kling M, Scroll E. Metallogenesis of germanium-a review. Ore Geology Reviews. 2007;**30**:145-180. DOI: 10.1016/j.oregeorev.2005.07.034

[12] Galley AG, Hannington MD, Jonasson IR. Volcanogenic massive sulfide deposits. In: Goodfellow WD, editors. Mineral Deposits of Canada: A Synthesis of Major Deposit-Types, District Metallogeny, the Evolution of Geological Provinces, and Exploration Methods. Special Publication; Vol. 5. Ottawa, Canada: Geological Association of Canada, Mineral Deposits Division; 2007. pp. 141-161

[13] Germanium. Webmineral Database [Internet]. Available from: http://webmineral.com/chem/Chem-Ge.shtml#.Wrj7i2ohLIU [Accessed: 02-03-2018]

[14] Radulescu M. Formation of a stratiform Zn-Pb-Ag SEDEX deposit—Numerical simulation. Carpathian Journal of Earth and Environmental Sciences. 2010;**5**:67-82

[15] Leach DL, Sangster DF, Kelley KD, Large RR, Garven G, Allen CR, Gutzmer J, Walters S. Sediment-hosted lead-zinc deposits: A global perspective Economic Geology. 100th Anniversary Volume. 2005. pp. 561-607

[16] Ye L, Cook NJ, Ciobanu CL, Yuping L, Qian Z, Tiegeng L, Wei G, Yulong Y, Danyushevskiy L. Trace and minor elements in sphalerite from base metal deposits in South

China—A LA-ICPMS study. Ore Geology Reviews. 2011;**39**:188-217. DOI: 10.1016/j.oregeorev.2011.03.001

[17] Kampunzu AB, Cailteux JLH, Kamona AF, Intiomale MM, Melcher F. Sediment-hosted Zn-Pb-Cu deposits in the Central African Copperbelt. Ore Geology Reviews. 2009;**35**:263-297. DOI: 10.1016/j.oregeorev.2009.02.003

[18] Hu R-Z, Qi H-W, Zhou M-F, Su W-C, Bi X-W, Peng J-T, Zhong H. Geological and geochemical constraints on the origin of the giant Lincang coal seam-hosted germanium deposit, Yunnan, SW China—A review. Ore Geology Reviews. 2009;**36**:221-234. DOI: 10.1016/j.oregeorev.2009.02.007

[19] Frenzel M, Ketris MP, Gutzmer J. On the geological availability of germanium. Mineralium Deposita. 2014;**49**:471-486. DOI: 10.1007/s00126-013-0506-z

[20] GERMANIUM U.S. Geological Survey. Mineral Commodity Summaries [Internet]. Available from: https://minerals.usgs.gov/minerals/pubs/commodity/germanium/mcs-2018-germa.pdf [Accessed: 12-03-2018]

[21] Li J, Zhuang X, Querol X, Font O, Izquierdo M, Wang ZM, Ketris MP, Gutzmer J. New data on mineralogy and geochemistry of high-Ge coals in the Yimin coalfield, Inner Mongolia, China. Journal of Coal Geology. 2014;**125**:10-21. DOI: 10.1016/j.coal.2014.01.006

[22] Frenzel M, Mikolajczak C, Reuter MA, Gutzmer J. Quantifying the relative availability of high-tech by-product metals—The cases of gallium, germanium and indium. Resources Policy. 2017;**52**:327-335. DOI: 10.1016/j.resourpol.2017.04.008

[23] Liu F, Liu Z, Li Y, Wilson BP, Lundström M. Extraction of Ga and Ge from zinc refinery residues in $H_2C_2O_4$ solutions containing H_2O_2. International Journal of Mineral Processing. 2017;**163**:14-23. DOI: 10.1016/j.minpro.2017.04.005

[24] Kul M, Topkaya Y. Recovery of germanium and other valuable metals from zin plant residues. Hydrometallurgy. 2008;**92**:87-94. DOI: 10.1016/j.hydromet.2007.11.004

[25] Zhang L, Xu Z. Application of vacuum reduction and chlorinated distillation to enrich and prepare pure germanium from coal fly ash. Journal of Hazardous Materials. 2017;**321**:18-27. DOI: 10.1016/j.jhazmat.2016.08.070

[26] Font O, Querol X, López-Soler A, Chimenos JM, Fernández AI, Burgos S, Peña FG. Ge extraction from gasification fly ash. Fuel. 2005;**84**:1385-1392. DOI: 10.1016/j.fuel.2004.06.041

[27] González A, Font O, Moreno N, Querol X, Arancibia N, Navia R. Smelting flue dust as a source of germanium. Waste and Biomass Valorization. 2017;**8**:2121-2129. DOI: 10.1007/s12649-016-9725-8

[28] Chen W-S, Chang B-C, Chiu K-L. Recovery of germanium from waste optical fibers by hydrometallurgical method. Journal of Environmental Chemical Engineering. 2017;**5**:5215-5221. DOI: 10.1016/j.jece.2017.09.048

[29] GERMANIUM U.S. Geological Survey. Mineral Commodity Summaries [Internet]. Available from: https://minerals.usgs.gov/minerals/pubs/commodity/germanium/mcs-2014-germa.pdf [Accessed: 02-02-2018]

[30] GERMANIUM U.S. Geological Survey. Mineral Commodity Summaries [Internet]. Available from: https://minerals.usgs.gov/minerals/pubs/commodity/germanium/mcs-2017-germa.pdf [Accessed: 02-02-2018]

[31] Bleiwas DI. Byproduct mineral commodities used for the production of photovoltaic cells: USGS 2010 [Internet]. Available from: http://pubs.er.usgs.gov/publication/cir1365 [Accessed: 02-11-2015]

[32] Weeks RA. Gallium, germanium, and indium. In: Brobst DA, Pratt WP, editors. United States Mineral Resources: Washington D.C., USA: US Geological Survey; Vol. 820. 1973. pp. 237-246

[33] Schwarz-Schampera U, Herzig PM. Indium Geology, Mineralogy, and Economics—A Review. Hannover, Germany: Berichte zur Lagerstätten- und Rohstofforschung, Bundesanstalt für Geowissenschaften und Rohstoffe; 1999. pp. 1-182

[34] GERMANIUM U.S. Geological Survey. Mineral Commodity Summaries [Internet]. Available from: https://minerals.usgs.gov/minerals/pubs/commodity/germanium/mcs-2016-germa.pdf [Accessed: 02-02-2018]

[35] Research on China Germanium Market, 2014-2018. Market Research Report [Internet]. Available from: https://www.radiantinsights.com/research/research-on-china-germanium-market-2014-2018 [Accessed: 12-03-2018]

[36] Germanium. CRM_InnoNet: Substitution of Critical Raw Materials [Internet]. Available from: http://www.criticalrawmaterials.eu/wp-content/uploads/Germanium- Citation-Style-Template-3.pdf [Accessed: 20-03-2018]

[37] UNEP-Recycling rates of metals—A Status Report, a Report of the Working Group on the Global Metal Flows to the international Resource Panel [Internet]. Available from: www.resourcepanel.org/file/381/download?token=he_rldvr [Accessed: 20-03-2018]

[38] Roskill Information Services Ltd, editor. Minor Metals in the CIS. 2nd ed. London: Roskill Information Services Ltd; 1997. p. 84

[39] Liu FP, Liu ZH, Li YH, Liu ZY, Li QH. Sulfuric leaching process of zinc powder replacement residue containing galium and germanium. The Chinese Journal of Nonferreous Metals. 2016;**26**:908-916

[40] Liu FP, Liu ZH, Li YH, Liu ZY, Li QH, Zeng L. Extraction of galium and germanium from zinc refinery residues by pressure acid leaching. Hydrometallurgy. 2016;**164**:313-320

[41] European Commission. Critical raw materials-European Commission [Internet]. Available from: http://ec.europa.eu/growth/sectors/raw-materials/specific-interest/critical_es [Accessed: 20-03-2018]

[42] Wardell MP, Davidson CF. Acid leaching extraction of Ga and Ge. The Journal of the Minerals, Metals and Materials Society. 1987;**39**:49-51

[43] Nusen S, Zhu Z, Chairuangsri T, Yong CC. Recovery of germanium from synthetic leach solution of zinc refinery residues by synergistic solvent extraction using LIX 63 and Ionquest 801. Hydrometallurgy. 2015;**151**:122-132

[44] Torma AE. Method of extracting gallium and germanium. Mineral Processing and Extractive Metallurgy Review. 1991;**7**:235-258

[45] Zhang LB, Guo WQ, Peng JH, Li J, Lin G, Yu X. Comparison of ultrasonic assisted and regular leaching of germanium from by-product of zinc metallurgy. Ultrasonics Sonochemistry. 2016;**31**:143-149

[46] Dutrizac JE, Chen TT, Longton RJ. The mineralogical deportment of germanium in the Clarksville electrolytic zinc plant of Savage Zinc Inc. Materials Transactions. 1996;**27**: 567-576

[47] Efremov VA, Potolokov VN, Nikolashin SV, Fedorov VA. Chemical equilibria in hydrolysis of germanium tetrachloride and arsenic trichloride. Inorganic Materials. 2002;**38**: 847-853

[48] Liang DQ, Wang JK, Wang YH. Difference in dissolution between germanium and zinc during the oxidative pressure leaching of sphalerite. Hydrometallurgy. 2009;**95**:5-7

[49] Lee HY, Kim SG, Oh JK. Process for recovery of gallium and germanium from zinc residues. Transactions of the Institution of Mining and Metallurgy. 1994;**103**:76-79

[50] Bauer DR, Cote G, Fossi P, Marchon B. Process for Selective Liquid–Liquid Extraction of Germanium; 1983. Patent US4389379 A

[51] Schepper AD. Liquid–liquid extraction of germanium by LIX 63. Hydrometallurgy. 1976;**1**:291-298

[52] Boateng DAD, Neudorf DA, Saleh VN. Recovery of Germanium from Aqueous Solutions by Solvent Extraction; 1990. Patent US4915919 A

[53] Schepper AD, Coussement M, Peteghem AV. Process for Separating Germanium from an Aqueous Solution by Means of an Alphahydroxyoxime; 1984. Patent US4432952 A

[54] Tian RC, Zou JY, Zhou LZ. New Technology for Indium, Germanium and Gallium Recovery in an Electrolytic Zinc Plant. Mineral Processing and Extractive Metallurgy. In: The International Conference Mineral Processing and Extractive Metallurgy. Kunming, China: Institution of Mining and Metallurgy; 1984. p. 615-624

[55] Zhou T, Zhong X, Zheng L. Recovering in, Ge and Ga from zinc residues. The Journal of the Minerals, Metals and Materials Society. 1989;**41**:36-40

[56] Wang H, Jiangshun L, Kaixi J, Dingfan Q. Recovery of Ga, Ge from zinc residues by hydrometallurgical processes. In: International Symposium; Advanced Processing of

Metals and Materials. New, Improved and Existing Technologies: Aqueous and Electrochemical Processing; September 2006. Vol. 6. pp. 413-420

[57] Ma X, Qin W, Wu X. Extraction of germanium (IV) from acid leaching solution with mixtures of P204 and TBP. Journal of Central South University of Technology. 2013; **20**:1978-1984

[58] Boateng DAD, Neudorf DA, Saleh VN. Recovery of Germanium from Aqueous Solutions by Solvent Extraction. 1989; Patent EP0313201 A1

[59] Energy Study, 2016 [Internet]. Available from: https://www.bgr.bund.de/EN/Themen/Energie/Produkte/energy_study_2016_summary_en.html [Accessed: 10-01-2018]

[60] Córdoba P, Font O, Izquierdo M, Querol X, Tobías A, López-Antón MA, Ochoa-Gonzalez R, Diaz-Somoano M, Martínez-Tarazona MR, Ayora C, Leiva C, Fernández PC, Gimenez AA. Enrichment of inorganic trace pollutants in re-circulated water streams from a wet limestone flue gas desulphurisation system in two coal power plants. Fuel Processing Technology. 2011;**92**:1764-1775

[61] Córdoba P. Partitioning and speciation of trace elements at two coal-fired power plants equipped with a wet limestone flue gas desulphurization (FGD) system [thesis]. Barcelona: Universitat Politènica de Catalunya; 2013

[62] Yao ZT, Ji XS, Sarker PK, Tang JH, Ge LQ, Xia MS, Xi YQ. A comprehensive review on the applications of coal fly ash. Earth-Science Reviews. 2015;**141**:105-121

[63] Ketris MP, Yudovich YE. Estimations of Clarkes for carbonaceous biolithes: World averages for trace element contents in black shales and coals. International Journal of Coal Geology. 2009;**78**:135-148

[64] Seredin VV, Finkelman RB. A review of the main genetic and geochemical types. International Journal of Coal Geology. 2008;**76**:253-289

[65] Calus-Moszko J, Bialecka B. Analysis of the possibilities of rare earth elements obtaining from coal and fly ash. Gospodarka Surowcami Mineralnymi-Mineral Resources Management. 2013;**29**:67-80

[66] Qin S, Sun Y, Li Y. Coal deposits as promising alternative sources for gallium. Earth-Science Reviews. 2015;**150**:95-101

[67] Nugteren H. Secondary Industrial Minerals from Coal Fly Ash and Aluminium Anodising Waste Solutions [thesis]. Delf: TU Delft; 2010

[68] Raask E. Mineral Impurities in Coal Combustion. Behaviour, Problems and Remedial Measures. Berlin, Germany: Springer-Verlag; 1985 ISBN 978-3540138174

[69] Font O. Extraction of potentially valuable elements [thesis]. Barcelona: Instituto de Ciencias de la Tierra Jaume Almera; 2007

[70] Waters RF, Kenworthy H. Extraction of germanium and gallium from coal fly ash and phosphorus furnace flue dust. Report 6940. US Bureau of Mines, 1967

[71] Zouboulis AI, Papadoyannis IN, Matis KA. Possibility of germanium recovery from fly ash. Chimiká chroniká New Series. 1998;**18**:87-97

[72] Hualing D, Zhide H. Ion flotation behaviour of thirty-one metal ions in mixed hydrochloric/nitric acid solution. Talanta. 1989;**36**:633-637

[73] Liang D, Wang J, Wang Y, Wang F, Jiang J. Behavior of tannins in germanium recovery by tannin process. Hydrometallurgy. 2008;**93**:140-142

[74] Marco-Lozar JP, Cazorla-Amorós D, Linares-Solano A. A new strategy for germanium adsorption on activated carbon by complex formation. Carbon. 2007;**45**:2519-2528

[75] Arroyo F, Fernandez-Pereira C. Hydrometallurgical recovery of germanium from coal gasification fly ash. Solvent extraction method. Industrial and Engineering Chemistry Research. 2008;**47**:3186-3191

[76] Hernández-Expósito A, Chimenos JM, Fernández AI, Font O, Querol X, Coca P, Garcia PF. Ion flotation of germanium from fly ash aqueous leachates. Chemical Engineering Journal. 2006;**116**:69-75

[77] Arroyo F, Font O, Fernandez-Pereira C, Querol X, Juan R, Ruiz C, Coca P. Ion flotation of germanium from fly ash aqueous leachates. Journal of Hazardous Materials. 2009;**167**:582-588

[78] Arroyo F, Font O, Chimenos JM, Fernández-Pereira C, Querol X, Coca P. IGCC fly ash valorisation. Optimisation of Ge and Ga recovery for an industrial application. Fuel Processing Technology. 2014;**124**:222-227

[79] Forsén O, Aromaa J, Lundström M. Primary copper smelter and refinery as a recycling plant-a system integrated approach to estimate secondary raw material tolerance. Recycling. 2017;**2**:1-19. DOI: 10.3390/recycling2040019

[80] Berry JB, Ferrada JJ, Dole LR, Ally M. Economic recovery of by-products in the mining industry [Internet]; 2001. Available from: https://info.ornl.gov/sites/publications/Files/Pub57727.pdf [Accessed: 11-02-2018]

[81] Font O, González A, Moreno N, Querol X, Navia R. Copper flash smelting flue dust as a source of germanium. Revista de la sociedad española de mineralogía macla. 2011;**15**;87-88

[82] Rentsch L, Aubel I, Schreiter N, Höck M, Bertau M. PhytoGerm: Extraction of germanium from biomass—An economic pre-feasibility study. Journal of Business Chemistry. 2016;**13**:47-58. DOI: 10.3390/recycling2040019

[83] Heilmeier H, Wilche O, Tesch S, Aubel IA, Schreiter N, Bertau M. Germaniumgewinnung aus Biomasse (PhytoGerm) [Internet]; 2010. Available from: http://www.vivis.de/phocadownload/Download/2016_rur/2016_RuR_177-192_Heilmeier.pdf [Accessed: 20-03-2018]

Chapter 3

Phosphorus and Gallium Diffusion in Ge Sublayer of $In_{0.01}Ga_{0.99}As/In_{0.56}Ga_{0.44}P/Ge$ Heterostructures

Kobeleva Svetlana Petrovna, Iliya Anfimov and
Sergey Yurchuk

Additional information is available at the end of the chapter

http://dx.doi.org/10.5772/intechopen.78347

Abstract

This chapter gives a short review on dopant diffusion in germanium and specifies the underlying mechanisms of diffusion that involve the point defects. Box-shaped diffusion profiles are discussed that may be described as the phosphorus diffusion controlled by doubly ionized vacancies. In this mechanism, the diffusion coefficient depends on the electron concentration. The particulars of P and Ga diffusion profiles in the Ga-doped substrate of $In_{0.01}Ga_{0.99}As/In_{0.56}Ga_{0.44}P/Ge$ heterostructures for multilayer solar cells are discussed. To calculate the diffusion coefficient, two methods were used: the Boltzmann-Matano (version of Sauer-Freise) and the coordinate-dependent diffusion analysis. It is established that coordinate-dependent diffusion analysis, which involves drift components together with diffusion components for diffusion profile description, is more suitable for description of the experimental profiles in such structures near p-n junction. A strong influence of intrinsic electric field on the dopant diffusivity was detected.

Keywords: P and Ga diffusion in Ge, A_3B_5/Ge heterostructures, box-shaped diffusion curve, impurity-vacancy complexes, coordinate-dependent diffusion method

1. Introduction

Impurity diffusion in semiconductors is one of the main processes for electronic device manufacturing, but on the other side, it could badly influence a semiconductor structure in multistage high-temperature electronic device manufacturing processes. Dopants, as phosphorus, at diffusion temperatures are ionized; therefore they actively interact with ionized lattice defects creating charged complexes. These complexes are formed and destroyed in the diffusion process that

leads to the appearance of generation and recombination components in a continuity equation that describes a diffusion process [1, 2].

Germanium is an important element to development of semiconductor theories and practice, and also it is a subject of many diffusion process researches. In this chapter, we focus on a narrow question: phosphorus diffusion in germanium, one of the main dopant of this material. Descriptions of diffusion processes were developing simultaneously with research of the crystalline and defect structure of this material and with improving of dislocation-free crystal growth technology together with development of measurement techniques and mathematical description of diffusion processes. That is why results that are 40 or 50 years old could be significantly different from contemporary ones. All these questions are under study and development. Progress in the first principal calculations together with the development of experimental techniques such as atomic force and scanning tunneling microscopy that allows to distinguish individual atoms and their lattice position will lead to the refinement of mechanisms and characteristics of diffusion processes. Our goal is to present the available data and knowledge about diffusion of phosphorus in germanium, possibly noting the problems and limitations of the representations used.

2. Phosphorus diffusion: first steps

Phosphorus, as a p-element of the group V of the periodic table, is a shallow donor impurity in germanium. The first works on phosphorus diffusion are about 1952–1954 years [3–5], and their review is in [2, 6].

It was previously mentioned that III and V group elements have a smaller diffusion coefficient than other groups of elements, and changes are mostly due to the frequency factor D_0. This was explained by their smaller ionic radius [5]. However, for elements of V group in germanium, this tendency was not confirmed (unlike that in silicon). Phosphorus, for example, having smaller ionic radius than any other V group element, has a smaller diffusion coefficient. For all shallow dopants (except of B), the activation energy is estimated as about 2.5 eV, and it slightly increased with decreasing diffusion coefficient in the range of As—Sb—P [5].

For a long time, constant diffusion coefficients were used for a fixed temperature [2–6]. These results were fairly expected, as in the absence of a reliable dopant profile measurement method, the diffusion coefficient was determined by p-n-junction depth; therefore it is in

$$D_P = 1.2 \cdot \exp\left(-\frac{2.5}{kT}\right) \mathrm{cm}^2 \cdot \mathrm{s}^{-1} \tag{1}$$

[5] and taking into account the semiempirical Langmuir-Dushman formula:

$$D_P = 2 \cdot \exp\left(-\frac{2.48}{kT}\right) \mathrm{cm}^2 \cdot \mathrm{s}^{-1} \tag{2}$$

At the same time, Ref. [5] already mentioned that high phosphorus concentration can lead to errors in calculations because of a tendency of this element to segregate. The surface

http://dx.doi.org/10.5772/intechopen.78347

concentration was not determined in the [5]. Another problem revealed in [5] was deviation of experimental values of p-n-junction depth in Sb diffusion (as the most studied dopant) from calculated dependence of p-n-junction depth on time ($d \sim \sqrt{t}$) at large time values. Therefore for estimation of the diffusion coefficient, a low diffusion time was used. Decrease of a penetration depth against expected one was attributed to diffusant evaporation in the diffusion process. These problems connected with the integral nature of a method of *D* coefficient determination.

In [7], the phosphorus profiles were determined using layered etching and sheet resistance measurements. Profiles of P in Ge that were made by vapor phase diffusion process were obtained for two surface phosphorus concentrations: less than and more than intrinsic carrier density n_i and at four diffusion temperatures—600, 650, 700, and 750°C. This allows to characterize temperature dependence of *D*. At low surface concentrations, the profile is described by Fick law, and diffusion coefficient is

$$D_1 = 330 \cdot \exp\left(-\frac{3.1}{kT}\right) \mathrm{cm}^2 \cdot \mathrm{s}^{-1} \tag{3}$$

At high surface concentration profiles which were extended, later [8] a name "box shaped" appears. For diffusivity calculations, authors applied Boltzmann-Matano method [1]. A dependence of the diffusion coefficient on the local phosphorus concentration was discussed. For the concentration-independent part, there was an expression obtained:

$$D_h = 0.01 \cdot \exp\left(-\frac{2.1}{kT}\right) \mathrm{cm}^2 \cdot \mathrm{s}^{-1} \tag{4}$$

Experimental data did not fit well into Arrhenius curves, especially for data at high phosphorus concentrations. With the temperature increase, the diffusion activation energy also increased.

Similar results were obtained in [8]. SIMS method was used for concentration profile measurements. Phosphorus diffusion was carried out at temperature range 600–910°C. Surface concentration of phosphorus was higher than 10^{19} cm^{-3}; therefore all samples were showing "box-shaped" profiles. Boltzmann-Matano method also was used for evaluating the concentration dependence of P diffusivity. The observed concentration dependence was approximately in agreement with results of [7]. The strong concentration dependence in *D* was attributed to dependence of *D* on Fermi level or due to strain effects caused by the difference in ionic radius of P in Ge.

In [8], temperature dependence of phosphorus diffusivity was found as

$$D_1 = (0.009 \pm 0.025) \cdot \exp\left(-\frac{2.1 \pm 0.2}{kT}\right) \mathrm{cm}^2 \cdot \mathrm{s}^{-1} \tag{5}$$

However the data of the paper allowed to derive another *D*(*T*):

$$D_1 = 1.21 \cdot \exp\left(-\frac{2.53 \pm 0.2}{kT}\right) \mathrm{cm}^2 \cdot \mathrm{s}^{-1} \tag{6}$$

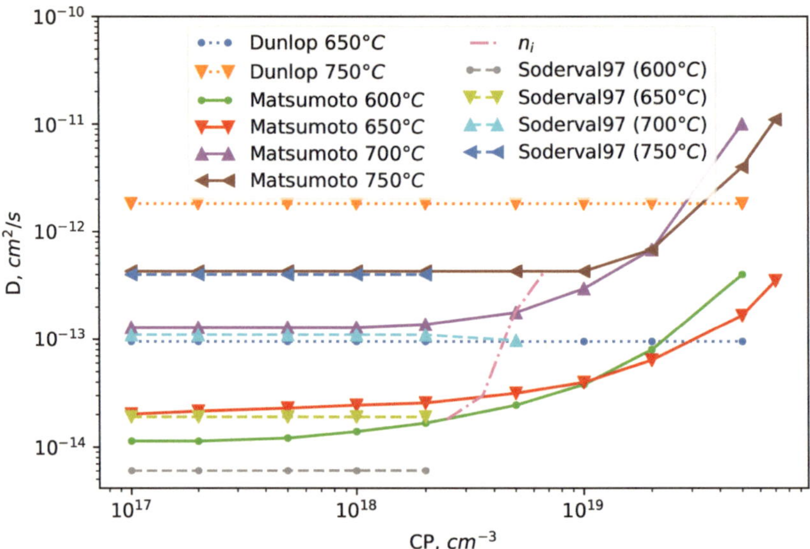

Figure 1. Diffusivity dependence on phosphorus concentration.

In the later works, a diffusion coefficient was called "intrinsic" for material, in which a dopant concentration $n < n_i$ at the growth temperature, and it was called "extrinsic" when $n > n_i$.

In **Figure 1**, there is the dependence of D on phosphorus concentration from [5, 7, 8]. Data of [8] were calculated by Eq. (6).

Integral values in [5] are noticeably higher than intrinsic D in [7, 8]; however, it does not exceed D in [7, 8] for high phosphorus concentrations.

Surprisingly, the experimental papers [7, 8] did not take into account extrinsic diffusion and dopant diffusion models, suggested in 1968 [9] and developed later [10–21]. Since vacancy in germanium is mostly acceptor with charge state up to −3, then positively charged phosphorus ion makes Coulomb-coupled pair with a charged vacancy. Diffusion of such pairs goes faster, and it was expected that it is in direct proportion to charged complex concentration.

3. Continuum theoretical calculations of dopant diffusion in semiconductors

The most detailed theory that describes dependence of dopant diffusivities on vacancy concentration in different charge states can be found in [10]. Indirect diffusion mechanisms, which involve vacancies V^k, are described by the following reaction:

$$(PV)^j + (j - m - k)e^- = P^m + V^k \tag{7}$$

The local equilibrium is characterized by

$$\frac{C_{P^m} \cdot C_{V^k}}{C_{(PV)^j} \cdot n^{j-n-k}} = const \tag{8}$$

where P^m-phosphorus in substitution position, V^k-vacancies in k-ionization state, $(PV)^j$—vacancy-phosphorus complex in j-ionization state.

Generally, reaction (7) is a fast process compared to time scale of diffusion, which typically amounts to several minutes up to several hours. For this condition local equilibrium of the reaction is reached.

For the conditions near equilibrium:

$$D_P = \frac{C_{(PV)^j} D_{(PV)^j}}{C_{P^m} + C_{(PV)^j}} \tag{9}$$

If $n \approx C_P > n_i$,

$$D^{eff}_{P^m_s} = (m+1) D_{(PV)^j} \left(C_{P^m_s}\right)^{m-j} \tag{10}$$

Thus, for $m = +1, j = -1, D^{eff}_P \sim D_{PV^j} \cdot n^2$, if $j = -2, D^{eff}_P \sim D_{PV^j} \cdot n^3$

In one dimension, the diffusion equation takes the form:

$$\frac{\partial C_x}{\partial t} + \frac{\partial J_x}{\partial x} = G_x, \tag{11}$$

where C_x and J_x, respectively, are the concentration and flux of point defect X ($P^m{}_s$, V^k, $P\text{-}V_j$) as a function of time t and position x. Possible reactions between X and other defects are taken into account by G_x. If flux is determined by the diffusion of X, that is,

$$J_x = -D_x \frac{\partial C_x}{\partial x} \tag{12}$$

The diffusion equation is given by

$$\frac{\partial C_x}{\partial t} - \frac{\partial}{\partial x}\left(D_x \frac{\partial C_x}{\partial x}\right) = G_x \tag{13}$$

In [10–21], the behavior of P and Sb was consistently explained by means of the double ionized vacancy mechanism:

$$PV^{-1} = P^{+1} + V^{2-} \tag{14}$$

If $m = +1, j = -1$, then $D^{\text{eff}}_P \sim n^2$

In [17], As, Sb, and P were used for diffusion experiments. A Ge-dopant alloy source with about 1 at. % dopant content was used. Diffusion anneals were performed at temperatures between 600 and 920°C for various times in vacuum. The multiple use of the dopant source leads to depletion of the source. So the maximum doping level could be changed from the values that exceed the intrinsic carrier concentration n_i to values close or beneath n_i at the diffusion temperature. Doping profiles with penetration depths in the range of 30–150 μm were measured by spreading resistance method. Secondary ion mass spectrometry was used to record diffusion profiles with depths of a few microns. It was confirmed that in the range of low dopant concentration, the intrinsic diffusion with the constant D_{in} has been occurred. The extra diffusion with "box-shaped" diffusion profiles was observed when dopant concentration exceeded n_i. In this case:

$$D^{eff}_{(PV)^-} = D_{(PV)^-}(n_i)\left(\frac{n}{n_i}\right)^2 \tag{15}$$

$$D_p(n_i) = 9.1^{+5.3}_{-3.4}\exp\left(-\frac{(2.85 \pm 0.04)eV}{k_B T}\right)cm^2 s^{-1} \tag{16}$$

Eq. (16) was calculated from Fickian-like profiles at low P concentrations. Then (15) were used for continuity equation, and a good agreement between experiment and calculations was achieved.

A "box-shaped" P profile was also detected under ion implantation procedure [18–21]. The "quadratic model" was used to describe diffusion process.

In [21], the phosphorus distribution in germanium after ion implantation and annealing at temperatures 523 and 700°C was measured by SIMS method. It was shown that neither quadratic nor constant diffusion coefficient models cannot be used for profiles at 700°C annealing and longtime annealing for both temperatures.

Later a cubic dependence of the P diffusivity on the electron concentration was proposed [22]. The equations and dependencies used were.

$$\frac{\partial C_P}{\partial t} = -\frac{\partial J_P}{\partial x} \quad J_P = -D^{eff} \cdot \frac{\partial C_P}{\partial x} - D^{eff} \cdot \frac{C_P}{n} \cdot \frac{\partial n}{\partial x} \tag{17}$$

$$D^{eff} = D^{2-}\left(\frac{n}{n_i}\right)^2 + D^{3-}\left(\frac{n}{n_i}\right)^3 \tag{18}$$

$$D_i = D^{2-} + D^{3-} \tag{19}$$

$$\begin{aligned} D_i &= 44.3 \cdot \exp\left(-\frac{3.01 \pm 0.04}{kT}\right)\mathrm{cm}^2 \cdot \mathrm{s}^{-1} \\ D^{2-} &= 11.1 \cdot \exp\left(-\frac{2.93 \pm 0.01}{kT}\right)\mathrm{cm}^2 \cdot \mathrm{s}^{-1} \\ D^{3-} &= 5.7 \cdot \exp\left(-\frac{2.92 \pm 0.02}{kT}\right)\mathrm{cm}^2 \cdot \mathrm{s}^{-1} \end{aligned} \tag{20}$$

There was a satisfactory conformity between experimental data and calculations for results of these authors and also with experimental data from [17] with this cubic model.

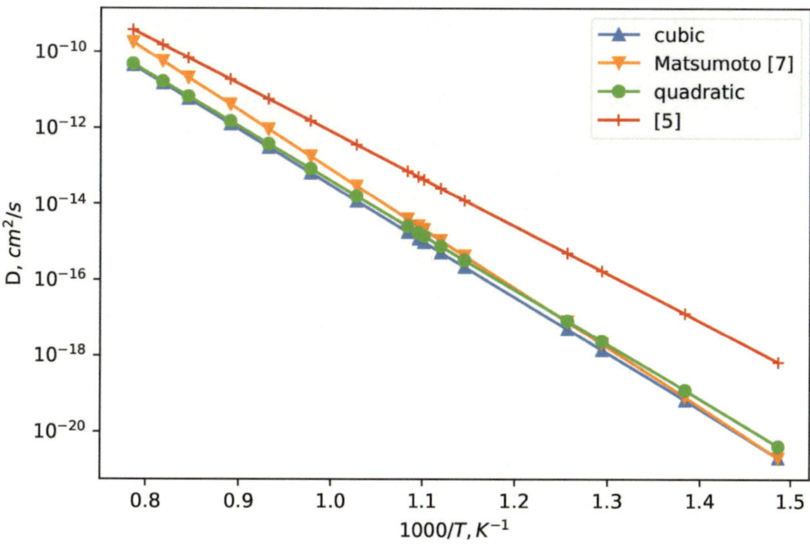

Figure 2. Intrinsic diffusivity for different models.

In **Figure 2**, a temperature dependence of the intrinsic diffusivity for cubic and quadratic models, experimental results in intrinsic diffusion regime [5] are presented. **Figure 3** demonstrates concentration dependence D for two models together with experimental dependence [5] if proposed $n = C_P$. As we can see, calculated by Boltzmann-Matano values of D differ from estimations of D_{in} from Fickian's part of diffusion curve, as it was done both in [17] for quadratic and [22] for cubic diffusion mechanisms.

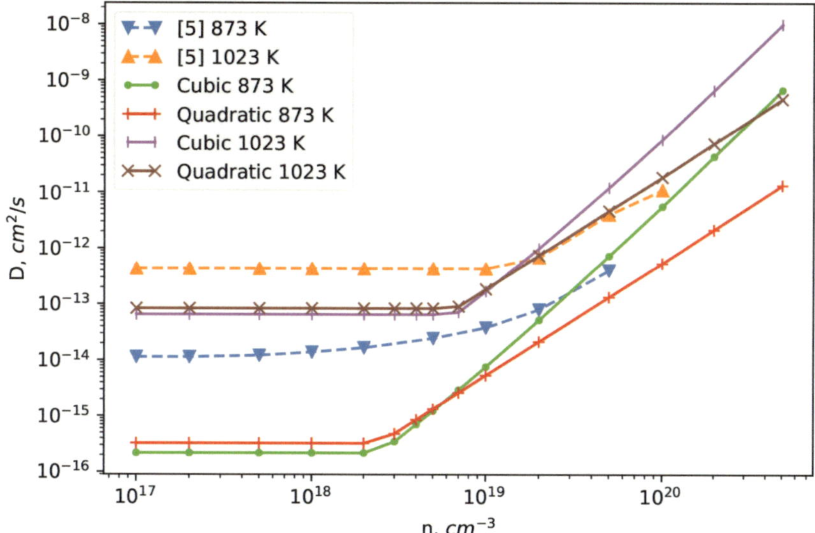

Figure 3. Dependencies of diffusivity for cubic and quadratic models. Dashed lines are from experimental results [7].

4. Diffusion of phosphorus in InGaAs/InGaP/P heterostructures

In [16] co-diffusion of Ga and P was investigated, and it was shown that co-doping strongly affects the diffusion of phosphorus. The interest to Ga and P co-diffusion appeared with the developments of multicascade solar cells.

In last two decades, germanium is considered as the most suitable material for the first cascade of multiple solar cells based on A^3B^5 compounds that is intended for transformation of the infrared solar spectrum [23]. Germanium cascade of the multiple solar cells is formed by phosphorus diffusion into heavily gallium-doped germanium substrates. It was found that p-n junction depth weakly depends on the diffusion time. In [24, 25], P and Ga profiles in the heterostructure $In_{0.01}Ga_{0.99}As/In_{0.56}Ga_{0.44}P/Ge$ were investigated. p-n junction of this element was formed at 635°C by phosphorus diffusion from $In_{0.56}Ga_{0.44}P$ buffer layer having thickness of about 24 nm to heavily doped of Ga germanium substrate ($C_{Ga} = 2{*}10^{18}$ cm^{-3}). The diffusion time was 2.6 min. SIMS has been applied to obtain profiles of P and Ga in heterostructure.

Figure 4 shows P, Ga, and free carrier concentration distribution in the Ge part of heterostructure. To calculate free electron concentration electroneutrality, equation was solved in the form of

$$C_P^+(x) + p(x) - n(x) - C_{Ga}^-(x) = 0 \tag{21}$$

As dopant concentrations near interface are high, Fermi-Dirac distribution was used [26]:

$$n = N_C \cdot F_{1/2}(\eta), p = N_V \cdot F_{1/2}(-\eta - \varepsilon_i) \tag{22}$$

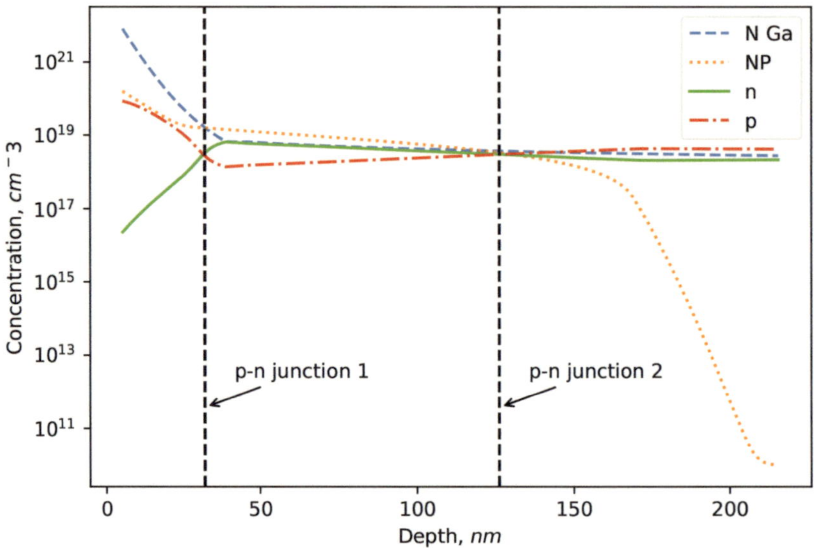

Figure 4. Profiles of P, Ga, n and p in Ge.

http://dx.doi.org/10.5772/intechopen.78347

where Fermi integral of order ½:

$$F_{1/2}(\eta) = \frac{2}{\sqrt{\pi}} \cdot \int_0^{\infty} \frac{\varepsilon^{1/2} d\varepsilon}{\varepsilon^{\varepsilon-\eta}+1}; \varepsilon = \frac{E-E_C}{kT}; \eta = \frac{F-E_C}{kT}; \varepsilon_i = \frac{E_C-E_V}{kT} \quad (23)$$

where F is the Fermi level and E_c and E_v are bottom of the conduction band and top of the valence band, respectively.

Numerical calculations of Fermi level were made by Newton method for defined concentrations of P and Ga.

It was found that Ga diffuses insensitive to Ge substrates together with P. The higher solubility of Ga than P was found on the InGaAs/Ge interface as it was also noted earlier [27] that leads to formation of two p-n junctions. The shallow p-n junction was formed at a depth of 30 nm and the second one at a depth of 130 nm. Diffusion part of Ga profile demonstrated Fickian-shaped curve with $D_{Ga} = 1.4 \times 10^{-15}$ cm²/s that exceeds data 6×10^{-17} - 2.3×10^{-16} cm²/s [4]. As it was expected, phosphorus profile has two parts: Fickian type near the surface in p-region ($C_{Ga} > C_P$) and box-shaped between p-n junctions where $n > n_i$. Unfortunately using diffusion coefficient with quadratic and cubic dependencies, the P profile could not be accurately described [25].

Two methods of diffusivity calculations were used [28]. The first one was Sauer-Freise (SF) method based on the Boltzmann-Matano calculation of diffusivity [1]. The second one was method of the analysis of coordinate-dependent diffusion (CDD) [29].

In the CDD method, two parameters are introduced that describe a probability of hopping process $\phi(x)$ and probability that the nearest vacant place for diffusion is empty $\gamma(x)$. Then diffusivity $D(x)$ and drift velocity $V(x)$ are expressed through these parameters and average distance between neighboring places λ. We have taken $\lambda = a = 0.566$ nm as a germanium lattice parameter. Then

$$D(X) = \phi(x)\gamma(x)\lambda^2 \quad (24)$$

$$V(x) = \left(\gamma(x)\frac{\partial\phi(x)}{\partial x} - \phi(x)\frac{\partial\gamma(x)}{\partial x}\right)\gamma^2 \quad (25)$$

Drift term includes continuity equation:

$$\frac{\partial C_X}{\partial t} - \frac{\partial}{\partial x}\left(D_X \cdot \frac{\partial C_X}{\partial x} - V(x)C_x\right) = 0 \quad (26)$$

Figure 5 shows calculated dependencies of P diffusivity on x for both methods. Positions of p-n junctions are presented. As we can see, diffusivity calculated using SF method is comparatively higher than using CDD method. That may be a consequence of existing a strong electric field in the sample in the p-n junction regions that leads to appearance of a strong drift component in the charged particle diffusion.

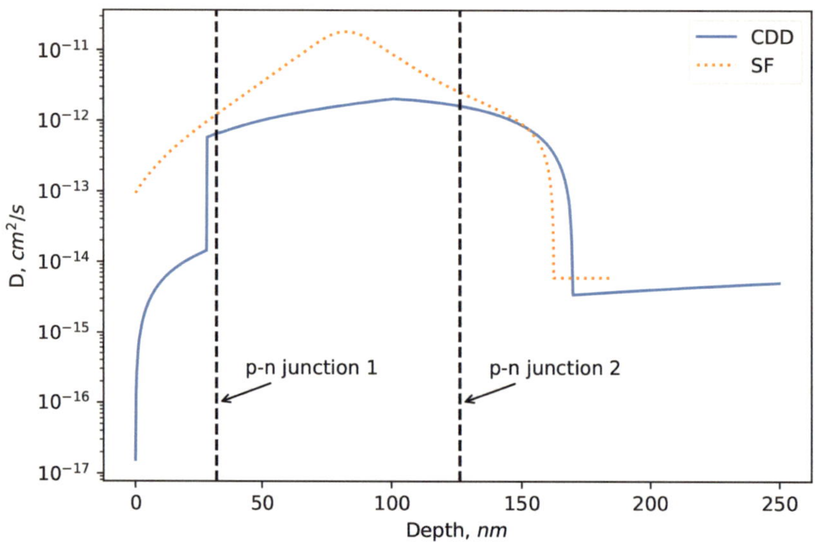

Figure 5. Diffusivity dependence on depth for $T = 635°C$.

Both methods of diffusivity calculations show two parts of D on x dependence: when x = 0–100 nm, diffusivity increases, and at higher values of x, diffusivity decreases. Width of the p side depletion region of the shallow left p-n junction on the **Figure 5** is of the order of 5–8 nm ($|C_{Ga} - C_P| < 10^{19}cm^{-3}$); both sides of right p-n junction in **Figure 5** are of the order of 50–80 nm ($C_{Ga} - C_p < 10^{17}cm^{-3}$); therefore an intrinsic electric field exists in the area between p-n junction. Approximately in the middle of junctions, the electric field changes its direction. Near the surface the intrinsic electric field accelerates negatively charged particles; when x > 100 nm, it inhibits diffusion. Outside of depletion regions (x > 160 nm), drift component of diffusion is negligible and diffusivity calculated by both methods which are equal.

Figure 6 shows dependencies of P diffusivity on n for Sauer-Freise, coordinate-dependent diffusion calculations, and different diffusion data from the literature.

An expected increase of the diffusivity with the free electron concentration was observed in both methods. Diffusivity produced by CDD calculation has two regions. The first one belongs to intrinsic diffusion ($n < n_i = 3.2 \times 10^{18}$ cm^{-3} [24]). As we can see, the lowest values of this part are equal to intrinsic diffusivity [5, 17, 30]. The second one corresponds to the diffusivity in the n-side of the p-n junctions and is higher than predicted both cubic and quadratic diffusion mechanisms. But the highest values of $D = 2 \times 10^{-12}$ cm^2/s at the $n = 7 \times 10^{18}$ cm^{-3} well correspond to maximum values [7], calculated by Boltzmann-Matano method. These values are observed in the electric field region of p-n-p structure that is formed in the germanium near the interface. Diffusivity dramatically drops at the ends of this structure in the p-region that may be connected with the shape of intrinsic electric field that in the case of linear p-n junction depend on x quadratically and drops sharply in the end of the depletion region. We can assume that the electric field causes not only the appearance of a drift component in diffusion but also increases the diffusivity of P-V pairs $D_{(PV)^j}$.

http://dx.doi.org/10.5772/intechopen.78347

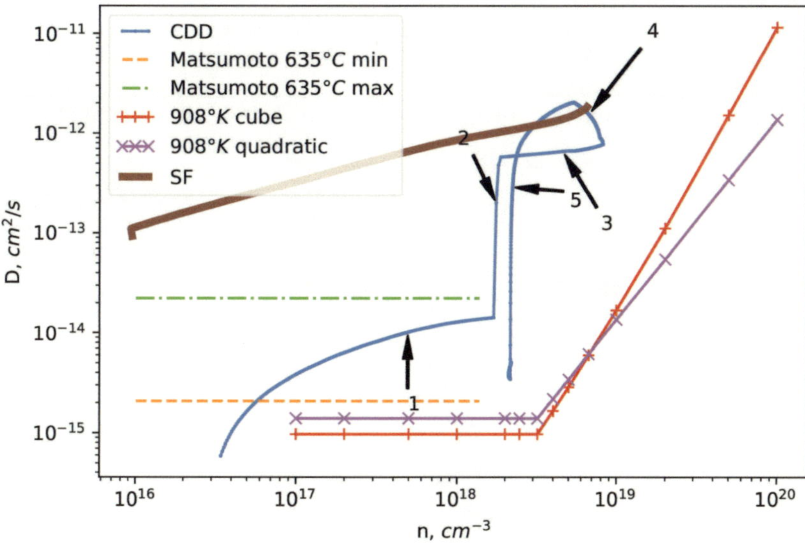

Figure 6. Diffusivity dependence on electron concentration for $T = 635°C$. 1: $0 < x < 25$ nm, 2: $25 < x < 33$ nm, 3: $33 < x < 60$ nm, 4: $60 < x < 100$ nm, 5: $x > 100$ nm.

There are two regions of weak dependence of D on n. The first one at $n < 2 \times 10^{18}$ cm^{-3} corresponds to intrinsic diffusivity and is quite expected. The second is observed at high n in the region where the electric field exists. To understand the weak dependence of D_P on n (3–4 on **Figure 6**), we shall consider the equations for P-V complexes forming. In **Table 1**, the equations and parameters k and j that lead to different dependencies of $D_{(PV)^j}$ on n (see (7)) are presented for two cases.

The first is the same as in [17] when $n = C_{P+}$; the second is for the case of C_{P+} = const as it is in our samples between p-n junctions (see **Figure 6**). We propose that $D_{(PV)^j} \sim C_{(PV)^j}$ [17].

Assumption that $n = C_P$ is valid in a material with one type of impurity. In a strongly compensated material, the concentration of free charge carriers is significantly lower than the concentration of the impurity. Between p-n junctions in the measured heterostructure, phosphorus concentration changes slowly and we may suggest C_{P+} = const. Phosphorus atoms in substitution positions are fully ionized, so m = +1; vacancies may be single, double, and triple ionized, that is, k = 0, −1, −2, and − 3, VP pairs—single and double ionized (j = 0, −1, −2).

Which type of reaction will be realized depends on the position of the Fermi level of the material, which controls the ratio of the centers in different charge state. The greater the electron concentration, the greater the charge state of acceptors, that is, for the condition $n = C_{P+}$, the most probable dependence of diffusivity of the complex is proportional n or n^2. In our case P-V complex should be charged, that is, j = −1 and −2. For C_{P+} = const a weak dependence of the diffusion coefficient on n is possible most likely for the reaction:

$$(PV)^{-1} = P^{+} + V^{-2} \tag{27}$$

	$C_P = n$			$C_P = const$		
	k	j		k	j	
n^0	−2	0	$(PV)^0 + e^- = P^+ + V^{-2}$	−1	0	$(PV)^0 = P^+ + V^{-1}$
	−3	−1	$(PV)^{-1} + e^- = P^+ + V^{-3}$	−2	−1	$(PV)^{-1} = P^+ + V^{-2}$
				−3	−2	$(PV)^{-2} = P^+ + V^{-3}$
n^1	−1	0	$(PV)^0 = P^+ + V$	0	0	$(PV)^0 = P^+ + V^0 + e^-$
	−2	−1	$(PV)^{-1} = P^+ + V^{-2}$	−1	−1	$(PV)^{-1} = P^+ + V^{-1} + e^-$
	−3	−2	$(PV)^{-2} = P^+ + V^{-3}$	−2	−2	$(PV)^{-2} = P^+ + V^{-2} + e^-$
n^2	0	0	$(PV)^0 = P^+ + V^{0^K} + e^-$	0	−1	$(PV)^{-1} = P^+ + V^0 + 2e^-$
	−1	−1	$(PV)^{-1} = P^+ + V^{-1} + e^-$	−1	−2	$(PV)^{-2} = P^+ + V^{-1} + 2e^-$
	−2	−2	$(PV)^{-2} = P^+ + V^{-2} + e^-$			
n^3	0	−1	$(PV)^{-1} = P^+ + V^0 + 2e^-$	0	−2	$(PV)^{-2} = P^+ + V^0 + 3e^-$
	−1	−2	$(PV)^{-2} = P^+ + V^{-1} + 2e^-$			

Table 1. Equations and parameters for different dependencies of $D_{(PV)^j}$ on n.

The ionization energies of different charge states must be known to estimate a charge of a defect. It is obvious that ionization energies of vacancies and vacancy-assisted complexities depend on the temperature, but there are no reliable data of that energies [15, 31–37]. In [36] it was shown that at equilibrium conditions, half occupancy of the doubly negatively charged state of the vacancy-group-V-impurity atom pairs occurs when the Fermi level is situated at the middle of the forbidden gap. In spite of large phosphorus concentrations, n in the case of our interest is comparatively small, Fermi level is near the middle of the forbidden gap, and we may suggest that the (27) is an achievement.

As the electron density increases, the charge state of the pair can change. In the depletion region of the first p-n junction together with sharp increase of the Fermi level, the amount and charge of the pairs can be changed drastically, leading to a sharp increase in D_P.

5. Conclusions

In spite of numerous P in Ge diffusivity investigations, there are some issues that remain unclarified. The first one is the discrepancies between intrinsic diffusivities, calculated from Fickean type of diffusion profile at low phosphorus concentrations and those calculated using Boltzmann-Matano method from diffusion profiles at high P concentration. If we agree with vacancy assistant diffusion model, it means that P introduction into Ge increases the total vacancy concentration.

The formation of a p-n junction for germanium cascade of multiple solar cells due to the diffusion of phosphorus from the buffer layer $In_{0.56}Ga_{0.44}P$ of $In_{0.01}Ga_{0.99}As/In_{0.56}Ga_{0.44}P/Ge$ heterostructure leads to co-diffusion of P and Ga. The process was held at 635°C for 2.6 min.

Solubility of Ga in the InGaP/Ge interface is higher than of P that leads to formation of two p-n junctions. Co-doping by gallium strongly affects the diffusion of phosphorus in germanium. We propose that it occurs primarily due to the electric field of the forming p-n junctions. P-type region is formed in the thin Ge surface layer (30 nm of order) with the depletion region thickness of 8–10 nm. The electric field of this p-n junction is directed to the Ge surface and accelerates both negatively charged Ga in interstitial positions and vacancy-phosphorus pairs. That leads to comparatively high gallium diffusivity $D_{Ga} = 1.4 \times 10^{-15}$ cm^2/s.

We can point out that in the case of Ga and P co-diffusion, calculations of diffusivity by Sauer-Freise and coordinate dependence diffusion methods give values an order of magnitude higher than the values, obtained for quadratic and cubic diffusion model for phosphorus diffusion. An electric field of a depletion region of p-n junctions leads to the appearance of drift components of phosphorus diffusion. At low electron concentrations in p-region near Ge surface in which there is no an electric field, phosphorus diffusivity increases with n from intrinsic diffusivity values, produced from Fickean-type profiles at low P concentration, to that one calculated by Boltzmann-Matano method for high P concentrations, while P concentration sharply decreases. We may suppose the vacancy concentration increasing as the concentration of Ga and P that occupied the vacancies decreased.

It can be assumed that the electric field causes not only the appearance of a drift component in diffusion but also increases the diffusivity of P-V pairs. The sharp diffusivity growth and drop are consistent with the electric field direction. In the first p-n junction, it is directed to the surface and accelerates negatively charged particles including Ga^- and $(PV)^-$. In the second one, it is directed into the sample that leads to decrease of the $D(PV)^-$.

For a correct description of the Ga and P co-diffusion, it is necessary to take into account both changes in the concentration of charged centers due to a change in the Fermi level position and the formation and decay of diffusing pairs. For this, in the continuity equation, it is necessary to take into account not only the drift component but also the generation-recombination terms corresponding to the formation and decomposition of the diffusing pairs.

Author details

Kobeleva Svetlana Petrovna*, Iliya Anfimov and Sergey Yurchuk

*Address all correspondence to: kob@misis.ru

National University of Science and Technology "MISiS", Moscow, RF, Russia

References

[1] Mehrer H. Diffusion in Solids. Fundamentals, Methods, Materials, Diffusion-Controlled Processes. Berlin, Heidelberg: Springer; 2007. 535 p

[2] Boltaks B. Diffusion in Semiconductors. New York: Academic Press; 1963. 462 p

[3] Dunlap W. Measurement of diffusion in germanium by means of pn junctions. Physics Review. 1952;**86**:615

[4] Dunlap W, Brown JR, Brown D. P-n junction method for measuring diffusion in germanium. Physics Review. 1952;**86**:417. DOI: 10.1103/PhysRev.86.417

[5] Dunlap W. Diffusion of impurities in germanium. Physics Review 1954;**94**:1531. DOI: https://doi.org/10.1103/PhysRev.94.1531

[6] Beke D, editor. Diffusion in Semiconductors and Non-metallic Solids. Subvolume A. Diffusion in Semiconductors. Berlin, Heidelberg: Springer; 1998. 575 p. DOI: 10.1007/b53031

[7] Södervall U, Friesel M. Diffusion of silicon and phosphorus into germanium as studied by secondary ion mass spectrometry. Defect and Diffusion Forum. 1997;**143-147**:1053-1058. DOI:10.4028/www.scientific.net/DDF.143-147.1053

[8] Matsumoto S, Niimi T. Concentration dependence of a diffusion coefficient at phosphorus diffusion in germanium. Journal of the Electrochemical Society. 1978;**125**:1307-1309

[9] Seeger A, Chik K. Diffusion mechanism and point defects in silicon and germanium. Physica Status Solidi (B). 1968;**29**:455-439. DOI: https://doi.org/10.1002/pssb.19680290202

[10] Bracht H. Self- and foreign-atom diffusion in semiconductor isotope heterostructures. I. Continuum theoretical calculations. Physical Review B. 2007;**75**:035210. DOI: https://doi.org/10.1103/PhysRevB.75.035210

[11] Shaw D. Self- and impurity diffusion in Ge and Si. Physica Status Solidi (B). 1975;**72**:11-39. DOI: https://doi.org/10.1002/pssb.2220720102

[12] Bracht H, Pedersen J, Zangenberg N, Larsen A, Haller EE, Lulli G, Posselt M. Radiation enhanced silicon self-diffusion and the silicon vacancy at high temperatures. Physical Review Letters. 2003;**91**:245502. DOI: https://doi.org/10.1103/PhysRevLett.91.245502

[13] Bracht H. Copper related diffusion phenomena in germanium and silicon. Materials Science in Semiconductor Processing. 2004;**7**:113-124. DOI: https://doi.org/10.1016/j.mssp.2004.06.001

[14] Bracht H, Brotzmann S. Atomic transport in germanium and the mechanism of arsenic diffusion. Materials Science in Semiconductor Processing. 2006;**9**:471-476. DOI: https://doi.org/10.1016/j.mssp.2006.08.041

[15] Brotzmann S, Bracht H, Hansen J, Larsen A, Simoen E, Haller E, Christensen J, Werner P. Diffusion and defect reactions between donors, C, and vacancies in Ge. I. Experimental results. Physical Review B. 2008;**77**:235207. DOI: https://doi.org/10.1103/PhysRevB.77.235207

[16] Naganawa M, Shimizu Y, Uematsu M, Itoh K, Sawano K, Shiraki Y, Haller E. Charge states of vacancies in germanium investigated by simultaneous observation of germanium self-diffusion and arsenic diffusion. Applied Physics Letters 2008;**93**:191905. DOI: https://doi.org/10.1063/1.3025892

[17] Brotzmann S, Bracht H. Intrinsic and extrinsic diffusion of phosphorus, arsenic, and antimony in germanium. Journal of Applied Physics. 2008;**103**:033508. DOI: https://doi.org/10.1063/1.2837103

[18] Chroneos A, Skarlatos D, Tsamis C, Christofi A, McPhail DS, Hung R. Implantation and diffusion of phosphorous in germanium. Materials Science in Semiconductor Processing. 2006;**9**:640-643. DOI: https://doi.org/10.1016/j.mssp.2006.10.001

[19] Tsouroutas P, Tsoukalas D, Zergioti I, Cherkashin N, Claverie A. Diffusion and activation of phosphorus in germanium. Materials Science in Semiconductor Processing. 2008;**11**: 372-377. DOI: https://doi.org/10.1016/j.mssp.2008.09.005

[20] Tsouroutas P, Tsoukalas D, Bracht H. Experiments and simulation on diffusion and activation of codoped with arsenic and phosphorous germanium. Journal of Applied Physics. 2010;**108**:024903. DOI: 10.1063/1.3456998

[21] Bracht H, Schneider S, Kube R. Diffusion, doping issues in germanium. Microelectronic Engineering. 2011;**88**:452-457. DOI: https://doi.org/10.1016/j.mee.2010.10.013

[22] Canneaux Th, Mathiot D, Ponpon J, Reques S, Schmitt S, Dubois Ch. Diffusion of phosphorus implanted in germanium. Materials Science & Engineering. B, Solid-State Materials for Advanced Technology. 2008;**154-55**:68-71. DOI: https://doi.org/10.1016/j.mseb.2008.08.004

[23] Kalyuzhnyy NA, Gudovskikh AS, Evstropov VV, Lantratov VM, Mintairov VM, Timoshina NKh, Shvarts MZ, Andreev VM. Germanium subcells for multijunction GaInP/GaInAs/Ge solar cells. Semiconductors. 2010;**44**:1520

[24] Kobeleva SP, Anfimov IM, Yurchuk SY, Turutin AV. Some aspects of phosphorus diffusion in germanium in $In_{0.01}Ga_{0.99}As/In_{0.56}Ga_{0.44}P/Ge$ heterostructures. Journal of Nano- and Electronic Physics. 2013;**5**(4):04001(3pp)

[25] Kobeleva SP, Anfimov IM, Yurchuk SYu, Vygovskaya EA, Zhalnin BV. Influence of $In_{0.56}Ga_{0.44}P/Ge$ heterostructure on diffusion of phosphor in germanium within the formation of multiple solar cells. Technical Physics Letters. 2013;**39**(1):27-29. DOI: 10.11 34/S1063785013010173/

[26] Zeeger K. Semiconductor Physics. Berlin Heidelberg: Springer; 2004. DOI: 10.1007/978-3-662-09855-4. 548 p

[27] Claeys C, Simoen E, editors. Germanium-based Technologies. From Materials to Devices. Oxford, Great Britain: Elsevier; 2007. 480 p

[28] Kobeleva SP, Anfimov IM, Yurchuk SY, Turutin AV. Influence of a Co-Doping on a Phosphorus and Gallium Diffusion in Germanium in $In_{0.01}Ga_{0.99}As/In_{0.56}Ga_{0.44}P/Ge$ Heterostructures. In: abstracts, XII International Conference on Nanostructured Materials. Lomonosov Moscow State University, Moscow, 13–18 July 2014, p. 685

[29] Malkovich R Sh. On the analysis of coordinate-dependent diffusion. Technical Physics. 2006;**51**(2):283-286. DOI:10.1134/S106378420602023X

[30] Canneaux T, Mathiot D, Ponpon J-P, Leroy Y. Modeling of phosphorus diffusion in Ge accounting for a cubic dependence of the diffusivity with the electron concentration. Thin Solid Films. 2010;**518**:2394-2397. DOI: https://doi.org/10.1016/j.tsf.2009.09.171

[31] Markevich V, Hawkins I, Peaker A, Emtsev K, Emtsev V, Litvinov V, Murin L, Dobaczewski L. Vacancy–group-V-impurity atom pairs in Ge crystals doped with P, As, Sb, and Bi. Physical Review B. 2004;**70**:235213

[32] Haesslein H, Sielemann R, Zistl C. Vacancies and self-interstitials in germanium observed by perturbed angular correlation spectroscopy. Physical Review Letters. 1998;**80**:2626

[33] Vainonen-Ahlgren E, Ahlgren T, Likonen J, Lehto S, Keinonen J, Li W, Haapamaa J. Identification of vacancy charge states in diffusion of arsenic in germanium. Applied Physics Letters. 2000;**77**:690. DOI: https://doi.org/10.1063/1.127087

[34] Fazzio A, Janotti A, da Silva A, Mota R. Microscopic picture of the single vacancy in germanium. Physical Review B. 2000;**61**:R2401

[35] Coutinho J, Öberg S, Torres V, Barroso M, Ones R, Briddon P. Donor-vacancy complexes in Ge: Cluster and supercell calculations. Physical Review B. 2006;**73**:235213

[36] Bracht H. Defect engineering in germanium. Physica Status Solidi A. 2014;**211**:109. DOI: 10.1002/pssa.201300151

[37] Dolidze ND, Tsekvava BE. On the model of divacancies in germanium. Physics of the Solid State. 2002;**44**:2034. DOI: https://doi.org/10.1134/1.1521452

Chapter 4

Interface Control Processes for Ni/Ge and Pd/Ge Schottky and Ohmic Contact Fabrication: Part One

Adrian Habanyama

Additional information is available at the end of the chapter

http://dx.doi.org/10.5772/intechopen.78692

Abstract

Metal-semiconductor interfaces are an essential part of any nano-electronic device. One of the concerns in germanium based technology is the presence of Fermi-level pinning (FLP) which leads to large Schottky barrier heights (SBH) for electrons. Details of the factors that pin the Fermi level will be discussed in this chapter. In an Ohmic contact there is an almost unimpeded transfer of majority carriers across the interface. One way to achieve such a contact is by doping the semiconductor heavily enough so that tunneling is possible. Heavy doping is not always advantageous or possible, depending on the type of device being fabricated. Other ways are to locally incorporate dopant atoms at the metal-germanium interface or to insert an interlayer into the interface. In practice, however, the contact resistivity is very sensitive to the interlayer thickness and the temperature of annealing used during the fabrication process. The latter two ways of achieving an Ohmic contact are interface control processes as opposed to the first way which is a bulk process. In this chapter we present the essential theoretical and experimental details required for the examination of some of the novel interface control processes developed for the fabrication of NiGe/*n*-Ge and PdGe/*n*-Ge Schottky and Ohmic contacts.

Keywords: thin film, Schottky barrier, Ohmic contact

1. Introduction

Germanium based nano-electronic technology suffers from two major limitations. In order to produce high speed devices, *n*-type germanium is preferred over *p*-type germanium because electrons have a higher mobility than holes. However, the doping levels in *n*-type germanium are low [1]. The second major limitation is that it is difficult to produce Ohmic contacts on *n*-type germanium [2–7] because of Femi-level pinning. One way to achieve Ohmic contacts is by

doping the semiconductor heavily enough so that tunneling is possible, this will be explained further in Section 1.1.2. However, heavy doping is a bulk process which is not always possible in *n*-type Ge. Other ways of producing Ohmic contacts are interface control processes like the local incorporation of dopant atoms at the metal-germanium interface or the insertion of an interlayer into the interface. The contact resistivity is very sensitive to the interlayer thickness and the temperature of annealing used during the fabrication process.

It has been demonstrated, in earlier studies [8], that NiGe/Ge and PdGe/Ge Schottky contacts have some of the lowest values of sheet resistivity in Ge-based technology. These contacts were also observed to remain stable over a wide temperature range during annealing [8, 9]. In this chapter we present the essential theoretical and experimental details required in order to make a comprehensive review of some of the interface control processes developed for the fabrication of NiGe/*n*-Ge and PdGe/*n*-Ge Schottky and Ohmic contacts; the review is presented in the next chapter.

1.1. Theory

Electrons in solids obey Fermi-Dirac statistics. At low temperatures, the distribution of electrons over a range of allowed energy levels at thermal equilibrium is given by,

$$f(E) = \frac{1}{1 + e^{(E-E_F)/k_B T}} \tag{1}$$

where, k_B is the Boltzmann constant. The function, $f(E)$ is the Fermi-Dirac distribution function which gives the probability that an available energy state, E will be occupied by an electron at temperature, T on the Kelvin scale. The quantity, E_F is the Fermi level. **Figure 1** shows a schematic illustration of the dependence of the Fermi-Dirac distribution function on electron energy at various temperatures.

At $T = 0$ K, $f(E) = 1$ for $E \leq E_F$ and $f(E) = 0$ for $E > E_F$. This means that there is a 100% probability that all available energy states, up to the energy, E_F, will be occupied at absolute zero, i.e., all energy levels up to E_F are occupied at 0 K. As the temperature is increased to T_1 and T_2 some energy levels which were occupied will become vacant and some energy levels, above the Fermi energy, which were vacant at absolute zero will become occupied. The probability, $f(E)$ at all temperature, T is equal to 0.5 when the energy E is equal to a quantity, μ called the chemical potential. At $T \approx 0$ K, $\mu = E_F$ and therefore $f(E_F) = 0.5$.

In applying the Fermi-Dirac distribution to semiconductors, we must recall that $f(E)$ is the probability of occupancy of an available state of energy, E. Thus, if there is no available electron state at the energy, E (e.g., if E is in the band gap of the semiconductor), there is no possibility of an electron having that energy. We can best visualize the relationship between $f(E)$ and the band structure of a semiconductor by turning the $f(E)$ versus E diagram on its side so that the E scale corresponds to the energies of the energy band diagram as shown in **Figure 2**.

Figure 2 represents intrinsic materials where the concentration of holes in the valence band is equal to the concentration of electrons in the conduction band and therefore the Fermi level E_F

http://dx.doi.org/10.5772/intechopen.78692

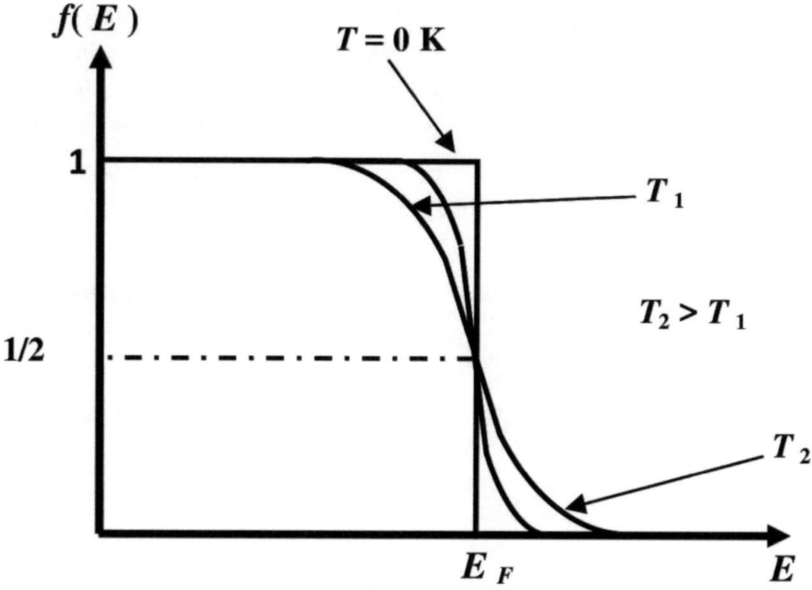

Figure 1. Schematic illustration of the dependence of the Fermi-Dirac distribution function on electron energy at various temperatures.

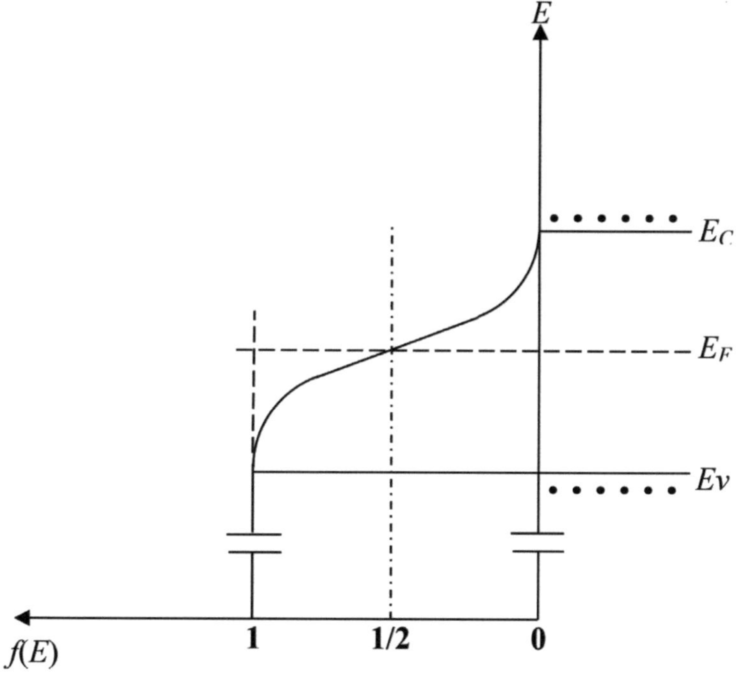

Figure 2. Visualization of the relationship between $f(E)$ and a semiconductor band structure by turning the $f(E)$ versus E diagram on its side so that the E scale corresponds to the energies of the energy band diagram.

lies near the middle of the band gap. In reality the effective densities of states, N_C and N_V in the conduction and valence bands respectively are slightly different because they depend on the effective inertial masses of the electrons and holes respectively, which are not the same. This causes the intrinsic Fermi level to be slightly displaced from the middle of the gap.

In *n*-type materials, there is a higher concentration of electrons in the conduction band than the hole concentration in the valence band. Thus, the Fermi level lies nearer the conduction band than the valence band, as shown in **Figure 3**.

In *p*-type materials there is a higher concentration of holes in the valence band compared with the electrons in the conduction band. The Fermi level therefore lies nearer the valence band than the conduction band, as seen in **Figure 4**.

In metals the valence and conduction bands overlap and there is not band gap. The Fermi level of a metal therefore lies in its conduction band, this fact will be referred to later on as we analyze **Figure 13** in Section 1.1.3 and **Figures 14** and **15** in Section 1.1.4.

1.1.1. Surface and interface states

Allowed electron energy states can be produced in the forbidden band gap of a semiconductor by the introduction of impurities or defects in the crystal. A metal-semiconductor interface introduces incomplete covalent bonds and other lattice defects at the semiconductor surface, which may result in the creation of interface states in the band gap. To explain a way in which

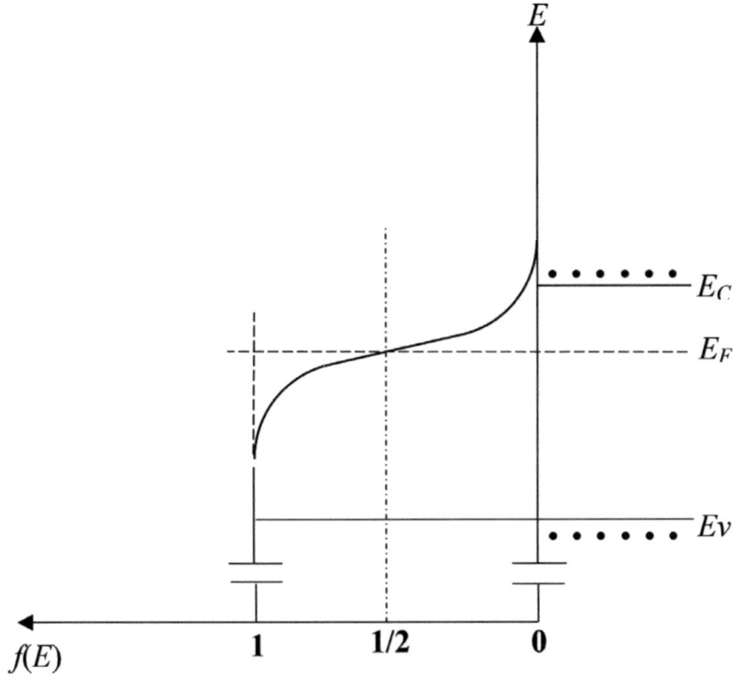

Figure 3. In *n*-type materials the Fermi level lies nearer the conduction band than the valence band.

http://dx.doi.org/10.5772/intechopen.78692

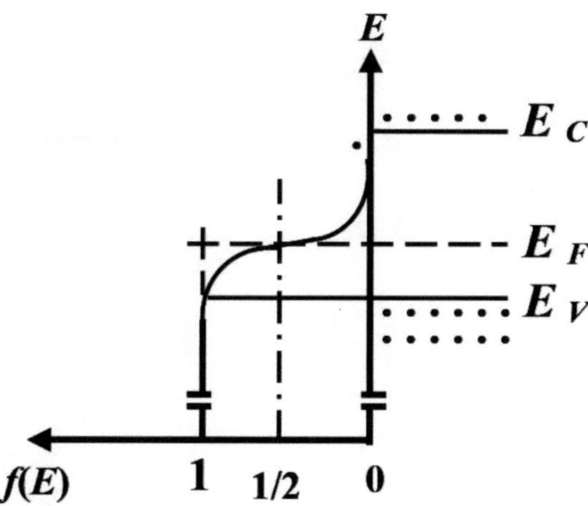

Figure 4. In *p*-type materials, the Fermi level lies nearer the valence band than the conduction band.

states may be created in the band gap, we could take the example of Ge doped with a donor impurity such as phosphorus (P) and an accepter like boron (B), as shown in **Figure 5**.

Phosphorus is in group V of the periodic table and is pentavalent. A P atom in the Ge lattice has the required number of valence electrons to complete the covalent bonding with four neighboring Ge atoms. The fifth valence electron of P does not fit into the bonding matrix of the Ge lattice and is therefore loosely bound to the P atom. Such electrons introduce energy levels very near the conduction band in the Ge band gap. These levels are occupied with electrons at 0 K and very little thermal energy is required to free them from the P atom, i.e., to excite them to the conduction band. At a temperature between 50 and 100 K, virtually all of the electrons in the impurity P levels are "donated" to the conduction band as shown in **Figure 6**.

Atoms like B from group III of the periodic table introduce accepter impurity levels in the Ge band gap near the valence band. B has only three valence electrons to contribute to the

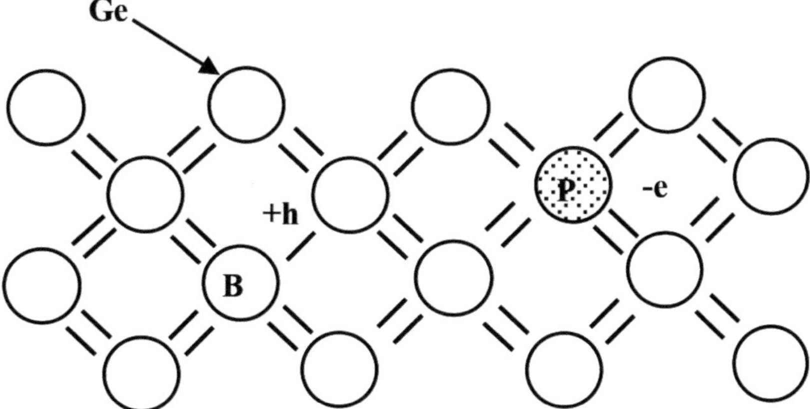

Figure 5. Schematic illustration of Ge doped with a donor impurity such as phosphorus (P) and an accepter like boron (B).

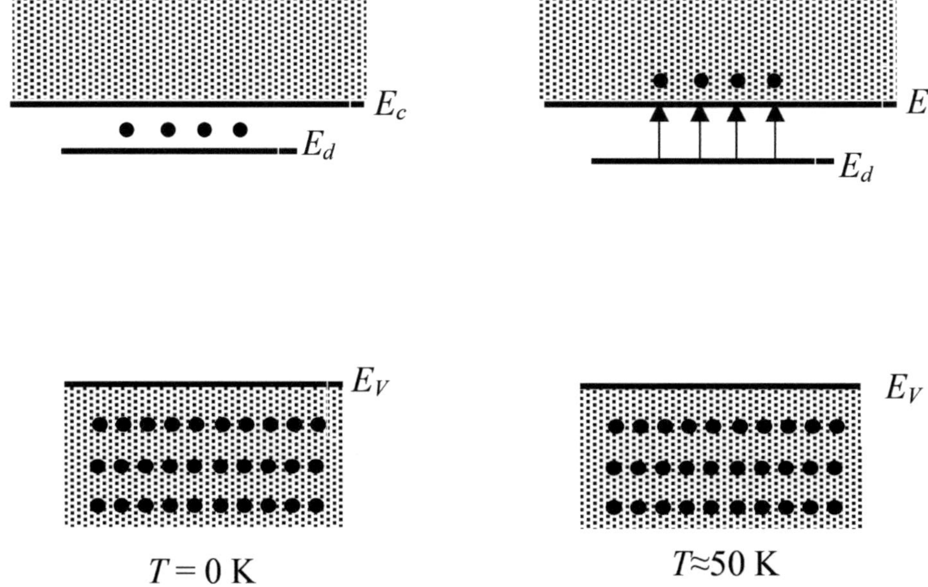

Figure 6. A donor level, E_d is occupied with electrons at 0 K and very little thermal energy is required to excite these electrons to the conduction band. Between 50 and100 K, virtually all of the electrons in the impurity level are "donated" to the conduction band.

covalent bonding thereby leaving one incomplete bond. Such levels are not occupied by any electrons at 0 K. As the temperature is increased the thermal energy increases enough to excite electrons from the valence band into the impurity levels, leaving holes in the valence band, which become current carriers when an external field is applied, by the continuous "hopping" of electron across adjacent incomplete bonds.

The incomplete bonds at a metal-semiconductor interface introduce energy levels in the band gap in a way similar to those due to accepter impurities. Interface traps, which introduce energy levels in the Ge band gap, can be caused by a sudden termination of a Ge crystal lattice at a metal/germanium interface.

1.1.2. Ideal Schottky barriers

A reference energy, E_0, called the vacuum energy, is the energy that an electron just "free" of a material would have in a vacuum. The work function of a semiconductor, Φ_s is defined as the energy required to move a unit electronic charge from the Fermi level to the vacuum level, i.e.,

$$\Phi_s = \frac{E_0 - E_{FS}}{q}, \tag{2}$$

where q is the electronic charge and E_{FS} represents the Fermi energy of the semiconductor. Similarly the work function of a metal is,

http://dx.doi.org/10.5772/intechopen.78692

$$\Phi_m = \frac{E_0 - E_{Fm}}{q}, \tag{3}$$

where E_{Fm} represents the Fermi energy of the metal. The electron affinity, χ_s of a semiconductor is defined as the energy required to move a unit electronic charge from the conduction band edge to the vacuum level, i.e.,

$$\chi_s = \frac{E_0 - E_C}{q}, \tag{4}$$

Figure 7 is a schematic diagram of the band structures of a metal and a semiconductor before contact for the case where $\Phi_m > \Phi_s$, the semiconductor Fermi level, E_{FS} is higher than that of the metal, E_{Fm}.

If $\Phi_m > \Phi_s$, the total energy of a metal/n-type semiconductor system could be reduced by moving electrons from the n-type semiconductor to the metal. When a metal is placed in contact with the semiconductor, therefore, electrons diffuse from the semiconductor to the metal in order to establish equilibrium. The electron diffusion causes the Fermi levels of the metal and the semiconductor to align at the same level throughout the interface region. Since the electron diffuse from the n-type semiconductor into the metal leaves behind uncompensated donor ions, a depletion region, W of an induced resultant positive charge is developed on the semiconductors side of the junction. A corresponding negative resultant charge is therefore induced on the metal

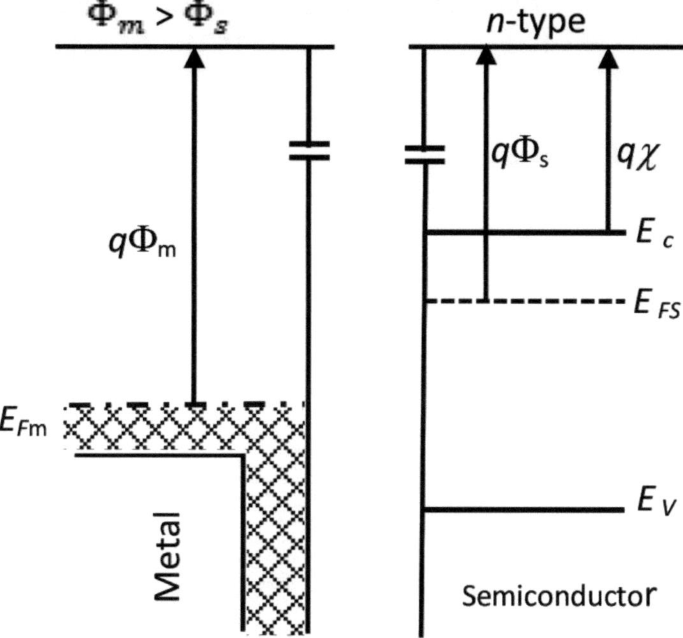

Figure 7. A schematic illustration of the band structures of a metal and a semiconductor before contact for the case where $\Phi_m > \Phi_s$, the semiconductor Fermi level, E_{FS} is higher than that of the metal, E_{Fm}.

side of the junction. The resultant positive charge from uncompensated donor ions in the depletion region matches the resultant negative charge induced on the metal, resulting in an electric field directed from the positive charge in the semiconductor to the negative charge in the metal. This causes the conduction energy band, *Ec* and the valence energy band, *Ev* of the semiconductor to bend in order to maintain continuity in the semiconductor band structure across the depletion region W, this is shown in **Figure 8**.

The electric field builds up to a magnitude where it eventually stops the electron diffusion across the junction, hence reaching a point of equilibrium. The corresponding induced equilibrium contact potential, V_o, across the junction, which prevents further electron diffusion from the semiconductor into the metal, is the difference in the work function potential energies, $\Phi_m - \Phi_s$, of the metal and the semiconductor, i.e., an energy of, $q(\Phi_m - \Phi_s)$ is required for an electron to cross from the semiconductor to the metal. The barrier V_o can be raised or lowered by the application of a voltage across the junction.

When a forward-bias voltage V is applied to the barrier the contact potential is reduced from V_o to $V_o - V$, as shown in **Figure 9**. As a result, electrons in the semiconductor's conduction

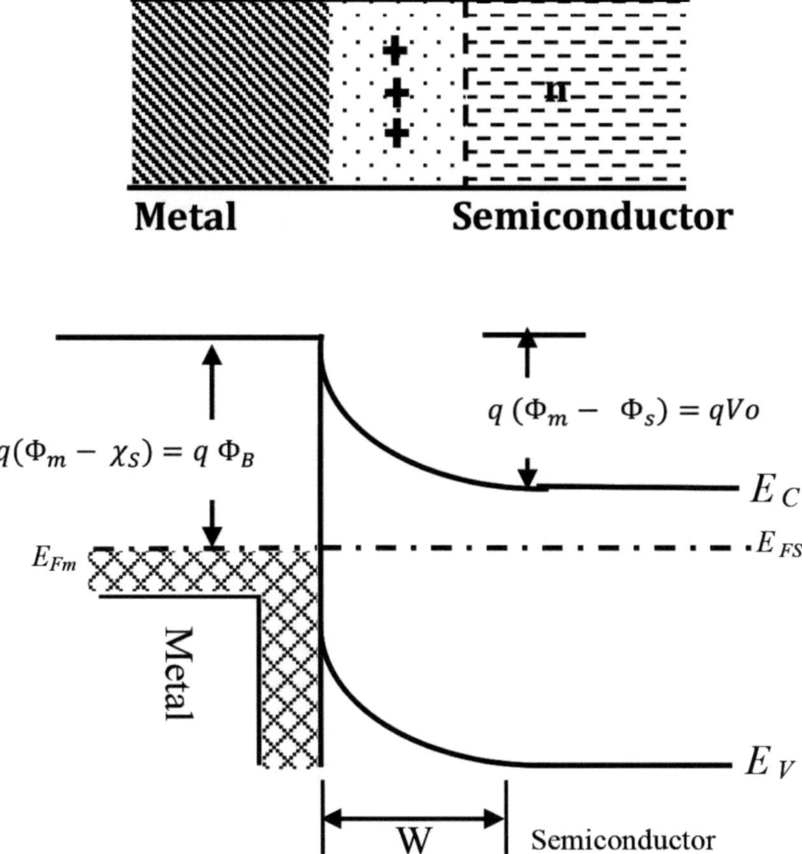

Figure 8. When a metal is placed in contact with a semiconductor the conduction, E_c and valence, E_v energy bands of the semiconductor bend in order to maintain continuity in the semiconductor band structure across the depletion region W.

http://dx.doi.org/10.5772/intechopen.78692

band can diffuse across the depletion region to the metal. This gives us the forward current from the metal to the semiconductor, since the direction of electron flow is opposite to the associated current direction.

When a reverse-bias voltage V_r is applied to the metal/*n*-type semiconductor junction, the contact potential is increased from V_o to a large potential barrier for electron flow from the semiconductor to the metal of, $V_o + V_r$ as shown in **Figure 10**. The electron flow from semiconductor to metal becomes negligible.

The reason that Ohmic contacts can be formed by heavy doping of the semiconductor, as mentioned in Section 1, is that even if this large, $V_o + V_r$ barrier exists at the interface, the heavy doping reduces the width of the depletion region, W to an extent that is small enough to allow electrons to tunnel through this barrier.

In both the forward and reverse-bias cases, electrons in the metal need to tunnel through an energy barrier of height,

$$q\Phi_B = q(\Phi_m - \chi_s), \tag{5}$$

in order to get into the semiconductor. The quantity, Φ_B, which will often also be labeled as Φ_{Bn} in this chapter, is referred to as the Schottky potential barrier height. This potential barrier

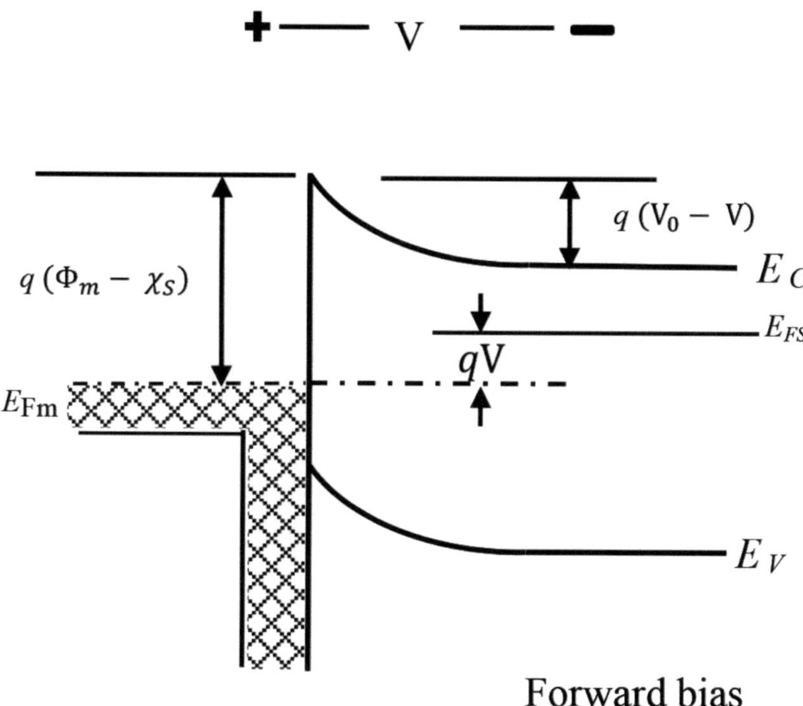

Figure 9. When a forward-bias voltage V is applied, the potential energy barrier from the semiconductor to the metal is reduced from V_o to $V_o - V$. Electrons are therefore able to tunnel across the barrier to give a forward current. Notice that the potential barrier from the metal to the semiconductor, $\Phi_m - \chi_S$ is much larger.

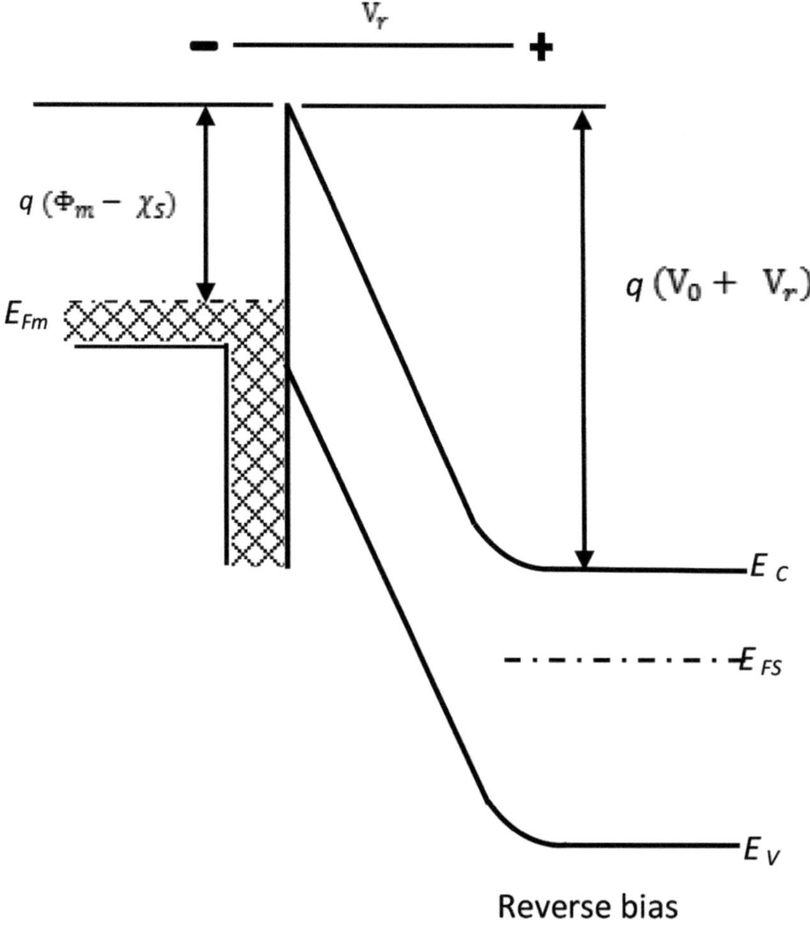

Figure 10. When a reverse-bias voltage V_r is applied, the barrier from the semiconductor to the metal is increased from V_o to $V_o + V_r$ and electron are not able to tunnel across the barrier. Notice that the potential barrier from the metal to the semiconductor remains, $\Phi_m - \chi_S$ as it was in the forward-bias case.

height is unaffected by the bias voltage but any reverse current due to electron injection from the metal into the semiconductor depends on the size of the Schottky potential barrier height, Φ_B. Electrons may flow from the metal to the semiconductor but the flow is retarded by the energy barrier, $q(\Phi_m - \chi_s)$. The contact therefore acts as a diode with I–V characteristics of the form sketched in **Figure 11**.

The resulting, I–V equation is similar in form to that of the *p-n* junction diode,

$$I = I_o\left(e^{qV/kT} - 1\right). \tag{6}$$

The reverse saturation current, I_0 depends on the size of the energy barrier, $q(\Phi_m - \chi_s)$ for electron injection from the metal into the semiconductor. This behavior of a metal/semiconductor contact is referred to as Schottky behavior as opposed to Ohmic behavior.

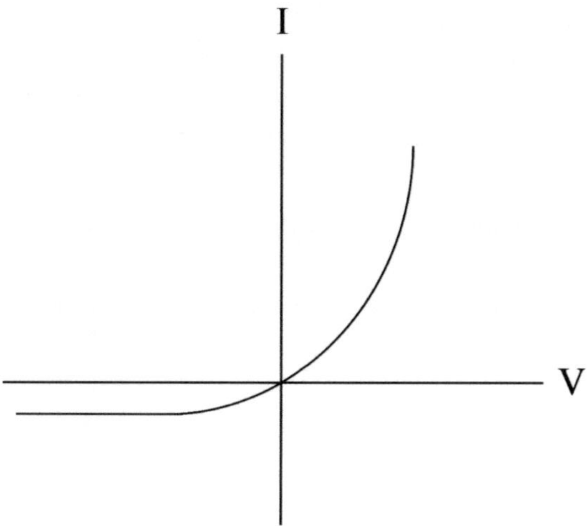

Figure 11. Schottky I–V characteristics where the reverse saturation current depends on the size of the energy barrier, $q(\Phi_m - \chi_s)$ for electron injection from the metal into the semiconductor.

1.1.3. Ohmic contacts

In many cases we wish to have an Ohmic metal/semiconductor contact, having linear I-V characteristics in both biasing directions. In *n*-type metal/semiconductor contacts, ideal (without Fermi level pinning) Ohmic behavior is observed if $\Phi_m < \Phi_s$. The separate energy bands for a metal and a semiconductor in this case are shown, before contact, in **Figure 12**.

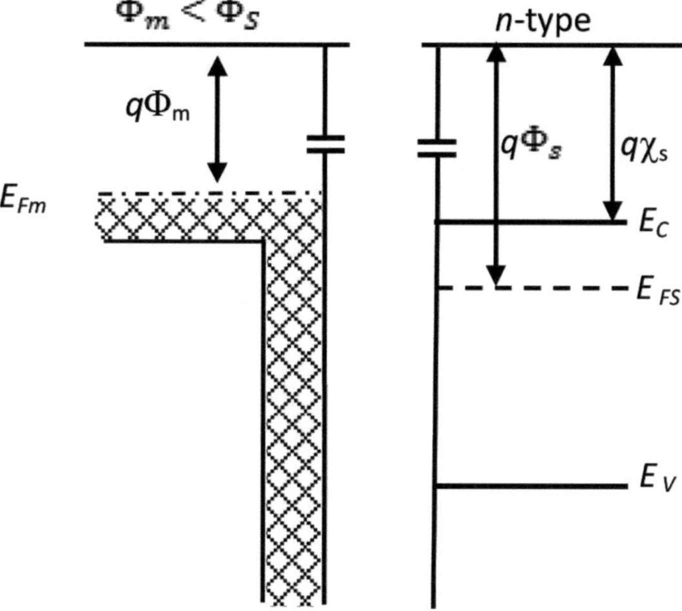

Figure 12. A schematic illustration of the band structures of a metal and a semiconductor, before contact, for the case where $\Phi_m < \Phi_s$. The semiconductor Fermi level, E_{FS} is lower than that of the metal, E_{Fm}.

After contact is made between the semiconductor and the metal, in the $\Phi_m < \Phi_s$ (*n*-type) case, the Fermi levels become aligned at equilibrium by transferring electrons from the metal to the semiconductor and not by the transfer of electrons from the semiconductor to the metal. The energy band structure, at the interface, for this case is illustrated in **Figure 13**.

We see in **Figure 13** that the conduction band of the semiconductor bends downwards towards the Fermi level of the metal at the interface, at equilibrium. What this means is that, since the Fermi level of a metal is in its conduction band as mentioned at the end of Section 1.1, the electrons in the metal are free to cross from their Fermi level straight into the conduction band of the semiconductor, i.e., electrons can flow unimpeded across the two conduction bands in both directions. Unlike in the rectifying contacts discussed earlier, no depletion region, W occurs in the semiconductor in this case.

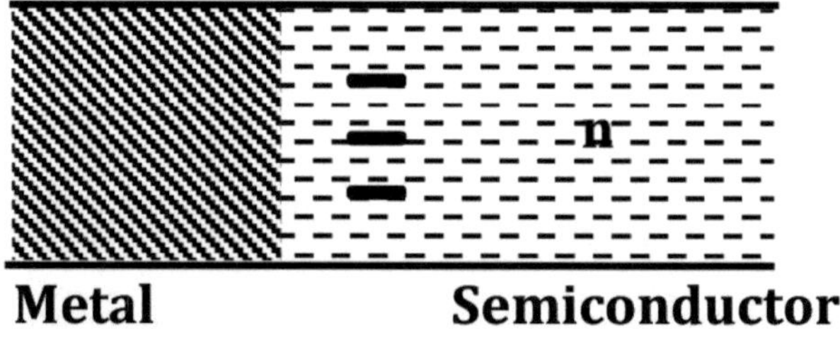

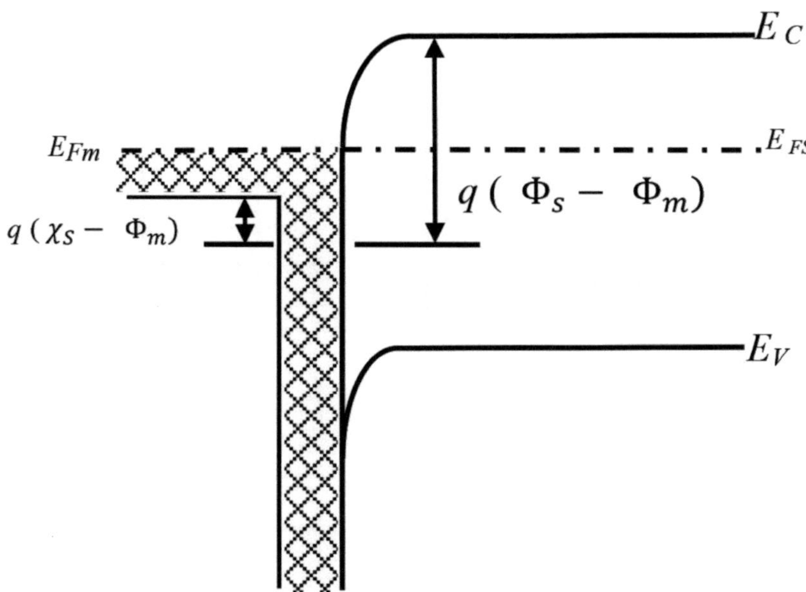

Figure 13. After contact is made between a metal and a semiconductor for the case where, $\Phi_m < \Phi_s$ the conduction band of the semiconductor bends downwards towards the Fermi level of the metal at the interface allowing electrons to flow unimpeded across the two conduction bands in both directions.

http://dx.doi.org/10.5772/intechopen.78692

1.1.4. *Practical Schottky barriers*

The ideal energy band diagram for a metal semiconductor interface has two main limiting factors. Firstly, the ideal contact does not take the surface states into account between a metal and a semiconductor. Secondly, when a practical metal/semiconductor interface is made, a thin interfacial layer is present on the semiconductor surface. This thin layer could potentially be a native oxide or processing residue, which contains a large density of surface states (D_{it}), many with energies distributed within the band gap of the semiconductor. The physics of the junction is then no longer governed by the properties of the metal and semiconductor materials alone but is then largely governed by the properties of the semiconductor surface [10].

The height of the Schottky potential barrier (Φ_B) is, in the ideal case, the difference between the metal work function (Φ_m) and the semiconductor's electron affinity (χ_s). A thin native oxide or processing residue insulating layer at the metal/semiconductor interface causes an additional voltage drop (V_i) over the metal/semiconductor interface, which is determined by the charge (Qs) at the semiconductor surface and the capacitance at the interface layer (C_i). Therefore,

$$\Phi_B = \Phi_m - \chi_s - V_i = \Phi_m - \chi_s - \left(\frac{Q_s}{C_i}\right). \tag{7}$$

It is possible to define a neutral level, Φ_0 in the interface energy band diagram. When the Fermi Level differs from the neutral level Φ_0, a net charge (Q_{it}) will be present at the semiconductor surface. Depending on the position of the surface states relative to Φ_0, the semiconductor surface will either be positive or negatively charged. For a very large density of surface states (D_{it}), the potential barrier height is only dependent on the band gap of the semiconductor and the neutral level of the semiconductor. Fermi level pinning then takes place at the interface making the potential barrier height independent from the metal work function. If the density of surface states is modeled as being infinitely large then the potential barrier height would be pinned at $(2/3)E_g$, which is known as Bardeen's limit [11]. The formation of surface states is dependent on the bonding type of the semiconductor material [10]. Covalent semiconductors such as germanium give rise to a large density of states at the surface due to the unsaturated bonds at the surface. For ionic semiconductors, the potential barrier height is more dependent on the metal work function [12]. In the following section we will define a quantity, n called the ideality factor and the symbol, Φ_{Bn} will be used for the Schottky potential barrier height.

The emission of electrons across a Schottky potential barrier can be described by two mechanisms: thermionic emission (TE) and diffusion. In practice, the transport process is a combination of both. The thermionic emission theory is based on a heat-induced flow of charge carriers from a surface over a potential energy barrier and is derived from the following assumptions [12, 13]:

1. The energy barrier height ($q\Phi_{Bn}$) is greater than the thermal energy of the electrons determined by k_BT, where k_B is Boltzmann's constant and T is the absolute temperature;
2. Thermal equilibrium is achieved at the plane that determines the emission;
3. Thermal equilibrium is not affected by the existence of a current flow. The two current fluxes, from the semiconductor to the metal and vice versa, can be superimposed;

4. The transfer of electrons across the interface of the metal and the semiconductor is the current limiting factor;

5. The electron mean-free-path should be bigger than the width of the region over, which a drop in potential energy, with a value of (k_B T), occurs at the barrier.

The total current density for thermionic emission, with an applied voltage, V across the barrier is given by [12, 14]:

$$J_T = J_{TO}\left[exp\left(\frac{qV}{nkT}\right) - 1\right], \tag{8}$$

where J_{TO} is the saturation current determined by

$$J_{TO} = A^*T^2 exp\left(\frac{-q\Phi_{Bn}}{kT}\right) \tag{9}$$

and

$$A^* = \frac{4\pi q m_n^* k^2}{h^3}. \tag{10}$$

The constant A^* is called the effective Richardson constant, q the electron charge, m_n^* the electron effective mass and h Planck's constant. The saturation current density (J_{TO}) is therefore independent of the applied voltage. Φ_{Bn} is the zero bias effective Schottky potential barrier height which can be obtained using the intercepts of the straight lines obtained by the extrapolation of J_{TO} in the semi-log forward bias ln J–V characteristics according to [15, 16]:

$$\Phi_{Bn} = \frac{kT}{q} ln\left(\frac{A^*T^2}{J_0}\right). \tag{11}$$

The factor n is equal to 1 for an ideal diode which conforms to pure thermionic emission but usually has values between 1 and 2. It determines the departure from the ideal diode characteristics and therefore modifies the diode equation, it is called the ideality factor, as mentioned earlier. The values of n can be calculated from the slopes of the linear regions of the semi-log forward bias ln J–V characteristics. It can be determined assuming pure thermionic emission [15, 16] using:

$$\frac{1}{n} = \frac{kT}{q}\frac{d}{dV}\ln(J). \tag{12}$$

The ideality factor is not a constant as it depends on the bias (V) and can only be specified for a particular point on a current-voltage characteristic curve.

The diffusion theory is based on the transport of charge carriers across a depletion region and is derived from the following assumptions [12]:

1. The energy barrier height ($q\Phi_{Bn}$) is greater than the thermal energy of the electrons determined by $k_B T$;
2. The effect of electron collisions taking place in the depletion region is included;
3. The current flow does not affect the carrier concentrations at the interface and in the semiconductor;
4. The impurity concentration of the semiconductor does not degenerate.

The diffusion current-voltage characteristics can be derived from the current density in the depletion region:

$$J_x = q\mu_n n(x)E(x) + qD_n \frac{\delta n}{\delta x}. \tag{13}$$

The diffusion current density in the x-direction depends on the electron charge, q the electron mobility, μ_n the electron concentration, $n(x)$, the electric field at the barrier, $E(x)$ and the diffusion coefficient for electrons, D_n. The current density can only be expressed in this form if the mobility and diffusion coefficient are independent of the electric field [10]. The total current density, J_D with an applied voltage across the barrier, V and temperature, T can be expressed, after applying Einstein's relationship ($Dn/\mu n = kT/q$), in the form:

$$J_D = J_{D0}\left[exp\left(\frac{qV}{kT}\right) - 1\right], \tag{14}$$

where the saturation current is,

$$J_{D0} = \left[\frac{q^2 D_n N_C}{kT}\left(\frac{2q(V_{bi} - V)N_D}{\varepsilon_{rs}\varepsilon_0}\right)^{1/2} exp\left(\frac{-q\Phi_{Bn}}{kT}\right)\right]. \tag{15}$$

The saturation current density, J_{D0} is determined by the effective density of states in the conduction band, N_C, the built in potential, V_{bi}, the donor concentration, N_D, the permittivity of free space, ε_0 and the relative permittivity of the semiconductor material, ε_{rs}.

We can see that the expressions for the current density are similar for the thermionic emission and diffusion theory and are based on the saturation current density. However, the saturation current density for the thermionic emission theory, J_{T0} is more sensitive to the temperature while the saturation current density of the diffusion theory, J_{D0} is more sensitive to the applied voltage [14]. It should be mentioned that a number of the results discussed in this chapter are for I–V as opposed to J–V characteristics, in such cases the equations presented in this section are applied with the incorporation of the perpendicular cross-sectional area of the current's flow.

If a Schottky diode is connected to electrodes which give a maximum electric field, E_m, there is what is referred to as an image-force which is the interaction due to the polarization of the conducting electrodes by the charged atoms of the sample. The image-force effect causes the energy barrier for electron transport across a metal-semiconductor interface to be lowered. The amount of barrier height reduction, $\Delta\Phi_B$ is given by [15, 16],

$$\Delta\Phi_B = \left[\frac{qE_m}{4\pi\varepsilon_{rs}\varepsilon_0}\right]^{1/2}. \tag{16}$$

In Schottky diodes, the depletion layer capacitance, C can be expressed as [15],

$$C^{-2} = \frac{2(V_0 - V)}{q\varepsilon_s A^2 N_D}, \tag{17}$$

where A is the cross-sectional area of the diode, V_0 is obtained from the intercept of the C^{-2}–V plot with the voltage axis and N_D is the donor concentration of the n-type semiconductor substrate. The value of N_D can be determined from the slope of the C^{-2}–V plot using Eq. (17). The maximum electric field E_m can be calculated using,

$$E_m = \left[\frac{2qN_D V_0}{\varepsilon_{rs}\varepsilon_0}\right]^{1/2}. \tag{18}$$

Due to the presence of surface states, an interfacial layer, microscopic clusters of metal-semiconductor phases and other effects, it is difficult to fabricate junctions with barriers near the ideal values predicted from the work functions and electron affinity. Therefore, measured barrier heights are used in device design and fabrication. In some semiconductors like Ge, the metal/semiconductor interface introduces states in the semiconductor band gap that pin the Fermi level at a fixed position, regardless of the metal used.

An example of a semiconductor Fermi level, E_{FS} that is pinned well below the conduction band edge is in n-type GaAs. In this case there is a collection of interface states located at energy position that are 0.7–0.9 eV below the conduction band. These state are responsible for pinning the Fermi level as explained later in the next section. The Fermi level, E_{FS} at the surface of the n-type GaAs is pinned at a position which is 0.8 eV below the conduction band edge, regardless of the choice of metal used, as shown in **Figure 14**. The Schottky barrier height is then determined from this pinning effect rather than by the work function of the metal. This means that electrons at the Fermi level of any metal in contact will always have to overcome the 0.8 eV barrier in order to cross over into the conduction band of the semiconductor.

A somewhat unique case of interest is in n-type InAs were E_{FS} at the interface is not pinned below but above the conduction band edge of the semiconductor, as shown in **Figure 15**. The semiconductor conduction band edge bends downwards at the interface with the metal just as illustrated for Ohmic contacts by **Figure 13** in Section 1.1.3. However, in this case, the bending of the semiconductor conduction energy band edge goes to a position, at the interface, which is below the Fermi levels of both the semiconductor E_{FS} and the metal E_{Fm}. Regardless of the metal in contact, the semiconductor Fermi level, E_{FS} remains in the conduction band (above E_C) of both the semiconductor and the metal. Since the Fermi level is the highest energy level filled with electrons at 0 K, this means that the electrons can freely cross from the metal to the semiconductor and back, at any temperature. Excellent Ohmic contacts to n-type InAs can therefore be produced by the deposition of almost any metal as a contact because of this Fermi level pinning in the conduction band.

http://dx.doi.org/10.5772/intechopen.78692

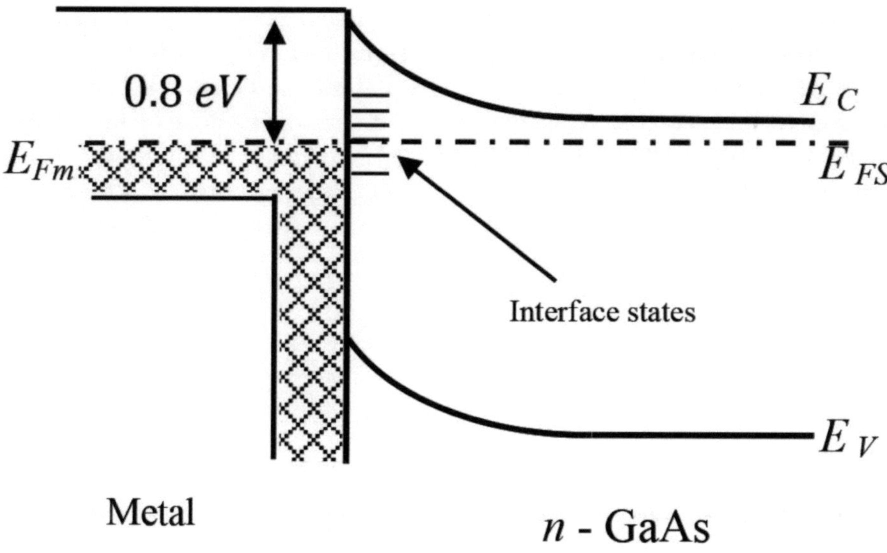

Figure 14. Notice that the semiconductor Fermi level, E_{FS}, lies 0.8 eV below the conduction band edge at the metal-semiconductor interface. This is regardless of the choice of metal used as a contact for GaAs, i.e., the Fermi level is pinned at this position.

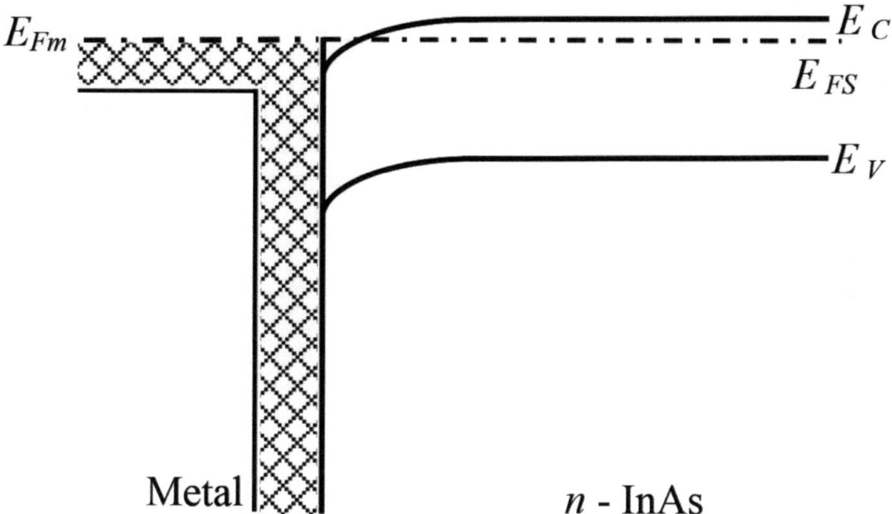

Figure 15. A schematic illustration of the case of *n*-type InAs. The semiconductor conduction band edge bends downwards at the interface with the metal, to a position below the Fermi level of the semiconductor, E_{FS}. This is regardless of the choice of metal used, therefore the Fermi level is pinned at this position above E_C.

1.1.5. Depinning of the interface

Fermi level pinning can be described by the theory of Metal Induced Gap States or MIGS [10]. In a metal/semiconductor junction the free electron wave function can penetrate into the semiconductor band gap. This generates band gap states, which consist of donor and acceptor like states. As mentioned in Section 1.1.4, there is a charge neutrality level, Φ_0 in the band gap where the gap-state charges are balanced. The Fermi level is pinned close to the charge neutrality level because of dipole formation. To prevent the Fermi level pinning, the free electron wave function penetration has to be reduced. This can be done by introducing a thin dielectric layer. Si_3N_4 [17–19] has low dielectric constant and moderately high band gap to prevent the free electron wave function from penetrating into the semiconductor band gap and hence releasing the Fermi level. Al_2O_3 [19] has also been reported to reduce the Fermi level pinning effects.

2. Experimental techniques

2.1. Metallization

There are principally three different methods of depositing metal on a substrate: plating, metal evaporation and sputtering. Metal plating is generally used to deposit thick layers. Only metal evaporation and sputtering were used in the work reported on in this chapter.

2.1.1. Metal evaporation

Evaporation techniques are based on heating up a source to a temperature where the material starts vaporizing. The vaporized material is then deposited on the sample and cools down forming a thin film. Thermal evaporation can either be achieved by heating the source with a resistive element or by using an electron beam. Resistive heating takes place by passing a current through a heating element, often made out of tungsten, which heats up a crucible containing the source material. Resistive evaporation has the disadvantage of potential contamination from the crucible if the melting temperature of the crucible is close to the melting temperature of the source material, resulting in a poor film quality. Electron beam evaporation uses an electron beam generated from a cathode to heat up the source material locally. The crucibles are water cooled to minimize contamination. The electron beam is generated by a thermionic emission filament and is accelerated towards the crucible using a high accelerating voltage. The beam is then focused into a spot on the surface of the source material and the interaction between the accelerated electrons and the source material will cause the material to start heating up and vaporize. The combination of local heating and water cooled sources prevents crucible metal contamination, resulting in a high purity film deposited on the substrate. The evaporation processes take place under high vacuum (10^{-3}–10^{-4} mTorr) in order to create a mean free path of the evaporating flux, which is greater than the distance between the source and the sample.

2.1.2. Sputtering

While evaporation requires a source to be heated to produce a flux of gas, sputtering targets make use of a physical plasma process rather than heat. The plasma is formed using an inert gas (normally Argon) and is excited by either a direct current (DC) or radio frequency (RF) source. The target source is negatively biased and the plasma sputters neutral atoms of source

http://dx.doi.org/10.5772/intechopen.78692

material away from the target towards an anode, where the neutral atoms are deposited on the sample. Since a plasma is required, the working pressures of sputtering systems are relatively high ($\approx 10^{-1}$ mTorr). The sputtering method used for most of the work reported on in this chapter is RF magnetron sputtering. Radio frequency magnetron sputtering is an enhanced sputter method which enables a higher deposition rate at low operating pressure together with the possibility of obtaining high quality films at low as well as high substrate temperatures. A schematic diagram of the experimental setup for this method is shown in **Figure 16**.

In the chamber filled with the Ar gas, a high voltage is applied at high frequency between the target and the sample. The surface atoms of the target material are removed and deposited onto the substrate by bombarding the target with the ionized Ar atoms. The magnet, located behind the target, enhances ionization and effectively directs the sputtered atoms towards the substrate.

2.1.3. *Cyclic stacking*

To explain the process of cyclic stacked we take the example of the production of an NiGe layer on a Ge substrate. Multi layers of Ni and Ge are formed by RF magnetron sputtering on an *n*-type germanium substrate at room temperature and the average composition of the whole multi-layers is controlled so as to have a stoichiometric equivalence to the atomic ratio of Ni

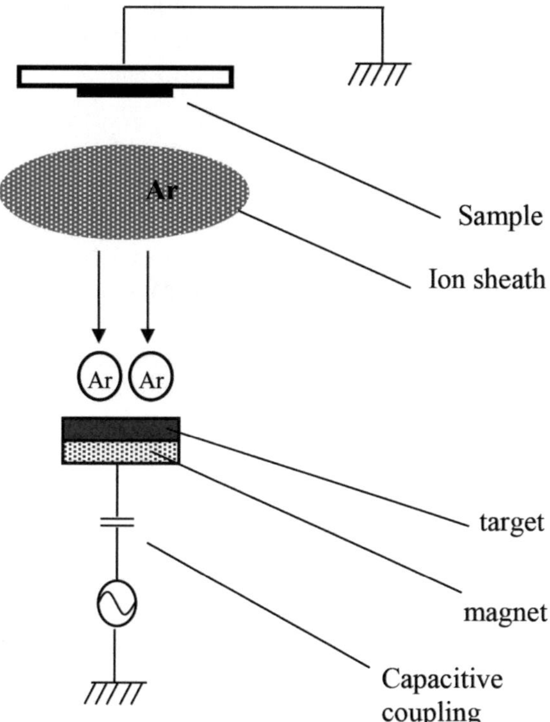

Figure 16. A schematic diagram of an RF magnetron sputtering unit.

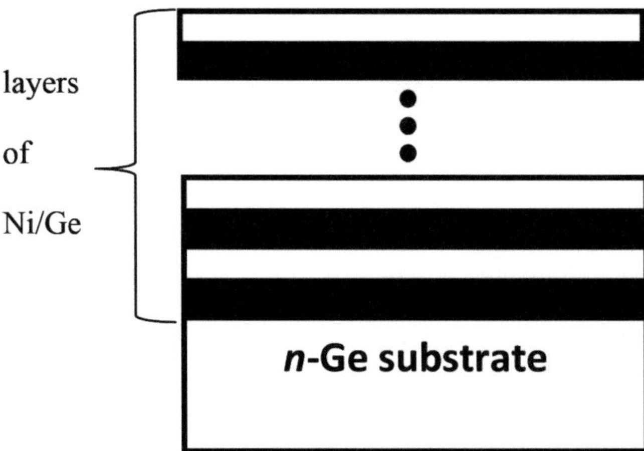

Figure 17. A schematic illustration of the sample configuration in cyclically stacked Ni/Ge films on an *n*-type Ge substrate.

and Ge atoms of 1 to 1 as in the phase, NiGe [20]. **Figure 17** is a schematic diagram showing the sample configuration in a cyclically stacked Ni/Ge film.

The idea behind this stacking of layers is to suppress the reaction between Ni and the Ge substrate upon annealing. In this way it is possible to get a high-quality NiGe film with a smooth interface on the Ge substrate. It is hoped that this smooth interface would reduce the Fermi level pinning effects of the interface electron energy states.

2.2. Donor implantation

Implanting atomic species like selenium (Se) into the surface of *n*-type germanium before metallization helps to reduce the Schottky barrier height by introducing local interfacial doping. In order to achieve a reasonable amount of implantation into the surface of the Ge substrate, the atoms to be implanted need to be energized to around 130 keV. The implantation is usually followed by heating at a high activation temperature to activate the diffusion of the dopant atoms further into the semiconductor surface, before metallization.

2.3. Four-terminal sheet resistivity measurement

The results of sheet resistance measurements presented in this chapter were obtained using a four-terminal resistor structure also known as a Kelvin resistor [21] structure. The structure consists of four contact pads: two pads are connected to the doped bulk semiconductor material and two pads contact to the metal used to form the contact. Current is then passed through two terminals between the semiconductor and the metal and the corresponding voltage drop is measured using the other two terminals between the metal and the semiconductor. In this way a sheet resistivity, ρ_{sh} can be extracted.

http://dx.doi.org/10.5772/intechopen.78692

Acknowledgements

The author would like to thank the Copperbelt University for the use of the institution's facilities.

Author details

Adrian Habanyama

Address all correspondence to: adrian.habanyama@cbu.ac.zm

Department of Physics, Copperbelt University, Kitwe, Zambia

References

[1] Simoen E, Schaekers M, Liu J, Luo J, Zhao C, Barla K, Collaert N. Defect engineering for shallow n-type junctions in germanium: Facts and fiction. Physica Status Solidi A: Applications and Materials Science. 2016;**213**(11):2799-2808

[2] Kittl JA, Opsomer K, Torregiani C, Demeurisse C, Mertens S, Brunco DP, Van Dal MJH, Lauwers A. Silicides and germanides for nano-CMOS applications. Materials Science and Engineering B. 2008;**154–155**:144-154

[3] Chawanda A, Nyamhere C, Auret FD, Mtangi W, Diale M, Nel JM. Thermal annealing behaviour of platinum, nickel and titanium Schottky barrier diodes on n-Ge (100). Journal of Alloys and Compounds. 2010;**492**(1-2):649-655. http://dx.doi.org/10.1016/j.jallcom.2009.11.202

[4] Martens K, Firrincieli A, Rooyackers R, Vincent B, Loo R, Locorotondo S, Rosseel E, Vandeweyer T, Hellings G, Jaeger BD, Meuris M, Favia P, Bender H, Douhard B, Delmotte J, Vandervorst W, Simoen E, Jurczak G, Wouters D, Kittl JA. Record low contact resistivity to n-type Ge for CMOS and memory applications. Technical Digest – International Electron Devices Meeting (IEDM). 2010. pp. 428-431

[5] Martens K, Rooyackers R, Firrincieli A, Vincent B, Loo R, De Jaeger B, Meuris M, Favia P, Bender H, Douhard B, Vandervorst W, Simoen E, Jurczak M, Wouters DJ, Kittl JA. Contact resistivity and fermi-level pinning in n-type Ge contacts with epitaxial Si-passivation. Applied Physics Letters. 2011;**98**:013504

[6] Firrincieli A, Martens K, Rooyackers R, Vincent B, Firrincieli A, Martens K, Rooyackers R, Vincent B, Rosseel E, Simoen E, Geypen J, Bender H, Claeys C, Kittl JA. Study of ohmic contacts to n-type Ge: Snowplow and laser activation. Applied Physics Letters. 2011;**99**: 242104

[7] Janardhanam V, Kim J-S, Moon K-W, Ahn K-S, Choi C-J. Annealing temperature dependency of the electrical and microstructural properties of Ti and Pt contacts to n-type Ge substrates. Microelectronic Engineering. 2012;**89**:10

[8] Gaudet S, Detavernier C, Kellock AJ, Desjardins P, Lavoie C. Thin film reaction of transition metals with germanium. Journal of Vacuum Science and Technology A. 2006;**24**:474

[9] Chawanda A, Nyamhere C, Auret FD, Mtangi W, Hlatshwayo T, Diale M, Nel JM. Thermal stability study of palladium and cobalt Schottky contacts on n-Ge (100) and defects introduced during contacts fabrication and annealing process. Physica B: Condensed Matter. 2009;**404**(22):4482-4484. https://www.sciencedirect.com/science/article/pii/S0921452609011156

[10] Rhoderick EH, Williams RH. Metal-Semiconductor Contacts. 2nd ed. Oxford: Oxford Science Publications, Clarendon Press; 1988

[11] Bardeen J. Surface states and rectification at a metal semi-conductor contact. Physical Review. 1947;**71**(10):717-727

[12] Sze SM. Physics of semiconductor devices. In: Colinge JP, Colinge CA, editors. Physics of Semiconductor Devices. New York: John Wiley and Sons; 1981. pp. 520-525

[13] Tiwari S. Compound Semiconductor Device Physics. London: Academic Press; 1992

[14] Troutman RR. VLSI limitations from drain-induced barrier lowering. IEEE Transactions on Electron Devices. 1979;**26**(4):461-469

[15] Sharma BL. Metal-Semiconductor Schottky Barrier Junctions and Their Applications. New York: Plenum Press; 1984

[16] Sze SM. Physics of Semiconductor Devices. 2nd ed. New York: Wiley; 1981

[17] Hu J, Guan X, Choi D, Harris JS, Saraswat K, Wong HSP. Fermi level depinning for the design of III-V FET source/drain contacts. In: International Symposium on VLSI Technology, Systems, and Applications, Hsinchu, Taiwan. Vol. 27-29. IEEE; . April 2009. pp. 123-124

[18] Kobayashi M. Fermi level depinning in metal/Ge Schottky junction for metal source/drain Ge metal-oxide-semiconductor field-effect-transistor application. Journal of Applied Physics. 2009;**105**(2):023702

[19] Hu J, Saraswat KC, Philip Wong HS. Metal/III-V Schottky barrier height tuning for the design of nonalloyed III-V field-effect transistor source/drain contacts. Journal of Applied Physics. 2010;**107**(6):063712

[20] Ishizaka A, Shirali Y. Solid-phase epitaxy of NiSi2 layer on Si(111) substrate from Si/Ni multi-layer structure prepared by molecular beam deposition. Surface Science. 1986;**174**(1-3):671-677

[21] Proctor SJ. A direct measurement of interfacial contact resistance. IEEE Electron Device Letters. October 1982;**3**(10):294-296

Chapter 5

Interface Control Processes for Ni/Ge and Pd/Ge Schottky and Ohmic Contact Fabrication: Part Two

Adrian Habanyama

Additional information is available at the end of the chapter

http://dx.doi.org/10.5772/intechopen.79318

Abstract

We examine the reported interface-based processes used in the modulation of Schottky barrier heights at the nickel germanide/*n*-type germanium and palladium germanide/*n*-type germanium junctions. Various sample preparation and characterization methods are discussed. Stable Ni/Ge and Pd/Ge structural phases are identified, and their temperature range of stability is established. Current-voltage (I-V) and capacitance-voltage (C-V) characteristics are analyzed to study the effect of various interface control processes. Sheet resistivity and its stability over various annealing temperature ranges are analyzed. The fundamental mechanisms at play in order to achieve ohmic characteristics are observed and analyzed using various interface control processes. Some interfacial and structural factors that pin the Fermi level are analyzed in relation to experimental results. The different interfacial control processes are analyzed, and their effectiveness is compared. Recommendations are made for the improvement of Ni and Pd contacts in the next generation of *n*-type germanium-based nanoelectronic devices.

Keywords: thin film, Schottky barrier, ohmic contact

1. Introduction

The fact that ohmic contacts provide an almost unimpeded transfer of majority carriers across an interface makes them an essential part of nanoelectronic device fabrication. The interface control processes of producing ohmic contacts in germanium-based technology, such as the local incorporation of dopant atoms at the metal-germanium interface and the insertion of an interlayer into the interface, result in contacts that have values of resistivity which are very sensitive to the interlayer thickness and the temperature of annealing used during the fabrication process. These aspects of the interface control processes will be examined in this chapter.

We present a review of some of the novel interface control processes developed for the fabrication of NiGe/*n*-Ge and PdGe/*n*-Ge Schottky and ohmic contacts.

2. Results and discussion

2.1. Phase-formation sequences

There has been a lot of work reported on the solid-state interactions in the Ni/Ge system [1–5] but interactions in the Pd/Ge system have not been as extensively reported on [6–8]. The available reports agree on the second and the final phase NiGe formed in the Ni/Ge system, but there is some disagreement on the first phase. There is agreement that Pd_2Ge is the first phase to be formed in the Pd/Ge system, the second and final phase to be formed is also agreed upon to be PdGe. These phases are generally reported to form sequentially [7, 9]. Our results [10, 11] for the phase-formation sequences, formation temperatures, and dominant diffusing species (DDS) during reactive diffusion in the Ni/Ge and Pd/Ge systems, obtained using in-situ (real-time) Rutherford Backscattering Spectrometry (RBS) and particle induced X-ray emission (PIXE) are summarized in **Table 1**.

We see from **Table 1** that in order to produce NiGe at an interface, the annealing temperature needs to be above 250°C. Below that temperature, Ni_5Ge_3 is produced. We also see that PdGe needs to be formed above an annealing temperature of 180°C, below which Pd_2Ge is formed. One positive aspect from these results in terms of device fabrication is that the two phases of interest, which are NiGe and PdGe, are the final phases to be formed in the Ni/Ge and Pd/Ge systems, respectively. What this means is that annealing at temperatures above 250°C and 180°C in the Ni/Ge and Pd/Ge systems respectively would effectively avoid the formation of the other phases of the systems.

Phases observed	Ni_5Ge_3, NiGe, Pd_2Ge, PdGe	
Phase-formation sequence	1st	Ni_5Ge_3, Pd_2Ge
	2nd	NiGe, PdGe
Phase-formation temperatures	Ni_5Ge_3	150°C
	Pd_2Ge	140–150°C
	NiGe	250°C
	PdGe	180°C
Diffusing species	Ni_5Ge_3	Ni
	NiGe	Ni is the DDS; Ge diffusion observed during the early stages of growth.
	Pd_2Ge	60% Pd and 40% Ge
	PdGe	65% Pd and 35% Ge

Table 1. Summary of our results for the thin film couple phase-formation sequences, phase-formation temperatures, and dominant diffusing species during the respective phase growths in the Ni/Ge and Pd/Ge systems [10, 11].

http://dx.doi.org/10.5772/intechopen.79318

2.2. NiGe contacts

2.2.1. Cyclically stacked NiGe contacts

One of the concerns regarding NiGe contacts on *n*-type Ge substrates is that other phases of the Ni/Ge system apart from NiGe are formed below 250°C. Another concern is the reaction of the deposited Ni film and the Ge substrate, which increases the interface roughness. Suppression of this interface reaction by the use of cyclic stacking, as explained in Section 2.1.3 of the previous chapter, is advantageous in obtaining a flat interface between NiGe and the Ge substrates. This was done in an investigation carried out by Motoki [12]. The wafers used in this study were *n*-type Ge (100) with a doping density of 4.0×10^{16} cm^{-3}. These substrates were treated with HF after which sets of Ni/Ge (0.5 nm/1.3 nm) layers were cyclically stacked eight times using RF magnetron sputtering. The thickness of the layers corresponded to an atomic ratio between Ni and Ge of 1 to 1, as in the phase NiGe. As explained in Section 2.1.3 of the previous chapter, the concept behind this process is to suppress the interface reaction, upon annealing, between the deposited Ni and the Ge substrate, hence reducing the number of interface electron energy states. The samples configuration is illustrated in **Figure 1**.

Two samples of cyclically stacked Ni/Ge were produced, one with 8 Ni/Ge cycles (referred to as sets in the figures) and the other with 16 cycles. In order to see if cyclic stacking produces improved results, two other samples were prepared with Ni films of thickness 3.0 and 5.5 nm respectively on Ge substrates without cyclic stacking, for comparison. The four types of samples, including the cyclically stacked ones, were annealed in nitrogen (N_2) gas at annealing temperatures that ranged from 200 to 500°C for 1 min. Four-terminal sheet resistance measurements were carried out on the samples as explained in Section 2.3 of the previous chapter. **Figure 2** shows experimental results of the sheet resistivity (ρ_{sh}) of the films as a function of the annealing temperature. We see a large decrease in sheet resistivity for the sample with a 3.0 nm-thick Ni film and no cyclic stacking within the temperature range from 200 to around 300°C. This is attributed to the formation of the NiGe phase. When the annealing temperature is over 350°C, the sheet resistivity shows a large increase owing to thermal instability. In the sample with a 5.5 nm-thick Ni layer and no cyclic stacking, the temperature range of the NiGe thermal phase stability is wider than that for the

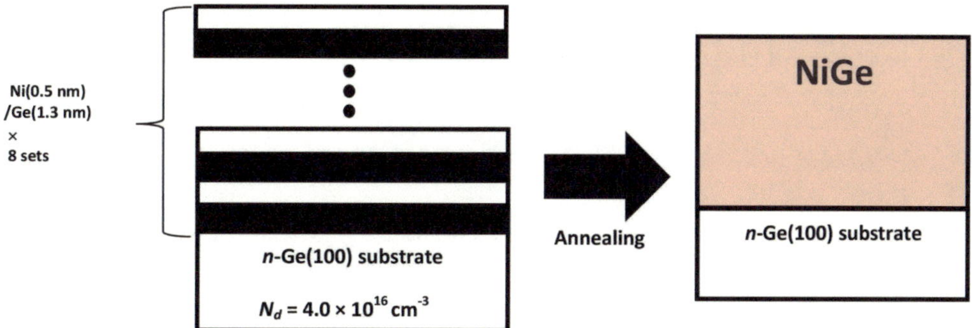

Figure 1. Cyclically stacked samples to suppress the interface reaction, upon annealing, between the deposited Ni and the Ge substrate.

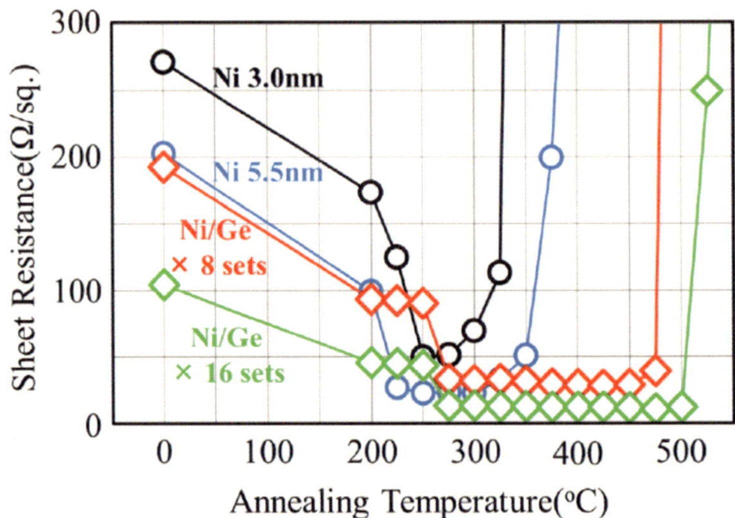

Figure 2. Experimental results of the sheet resistivity (ρ_{sh}) of Ni/Ge films as a function of the annealing temperature [12].

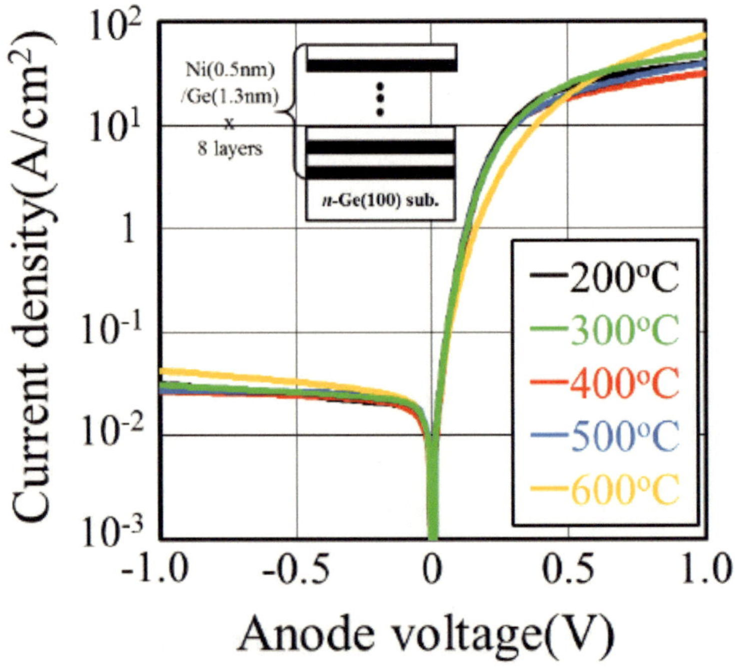

Figure 3. Current-voltage characteristics of the cyclically stacked NiGe at various annealing temperature from 200 to 600°C [12].

sample with the 3.0 nm-thick Ni layer. On the other hand, for the samples with cyclic stacking, a reduction in ρ_{sh} was observed over 250°C, and the value became stable at annealing temperatures from 275 to around 500°C. This demonstrates that cyclic stacking produces improved results.

Figure 3 shows the current-voltage characteristics of the cyclically stacked NiGe at various annealing temperature from 200 to 600°C. It is seen that the current density profile on a

http://dx.doi.org/10.5772/intechopen.79318

semilogarithmic scale in the reverse bias directions (negative anode voltage region) are very small compared to those in the forward bias direction, showing that the contacts are rectifying, which is typical Schottky diode behavior.

The height of the Schottky potential barrier, Φ_{Bn} and the ideality n-factor were extracted from the I-V characteristics of the Schottky diodes at various annealing temperatures using the thermionic emission model, as explained in Section 1.1.4 of the previous chapter. Equations (11) and (12), of the previous chapter, were used to extract Φ_{Bn} and the ideality n-factor respectively. The value of the effective Richardson constant, A^* used in this study was 133 A/cm^2 K^2. The results for the sample with a 5.5 nm-thick Ni layer (no cyclic stacking) and the Ni/Ge cyclically stacked sample with eight layers are shown in **Figure 4**, for annealing temperatures up to 600°C.

It is seen in **Figure 4** that the determination of both Φ_{Bn} and the ideality n-factor was repeated a number of times at each temperature, showing some experimental scattering errors, but the general trends are clear. The values of Φ_{Bn} that are determined for the sample with a 5.5 nm Ni film and the cyclically stacked Ni/Ge sample, annealed up to 600°C were within 0.54–0.57 and 0.53–0.55 eV, respectively. The ideality factor showed values of less than 1.3 for the cyclically stacked sample. We see in **Figure 4** that in this sample, the ideality factor could be maintained at values lower than 1.2 up to a temperature of 500°C, and it increases slightly at 600°C. In the

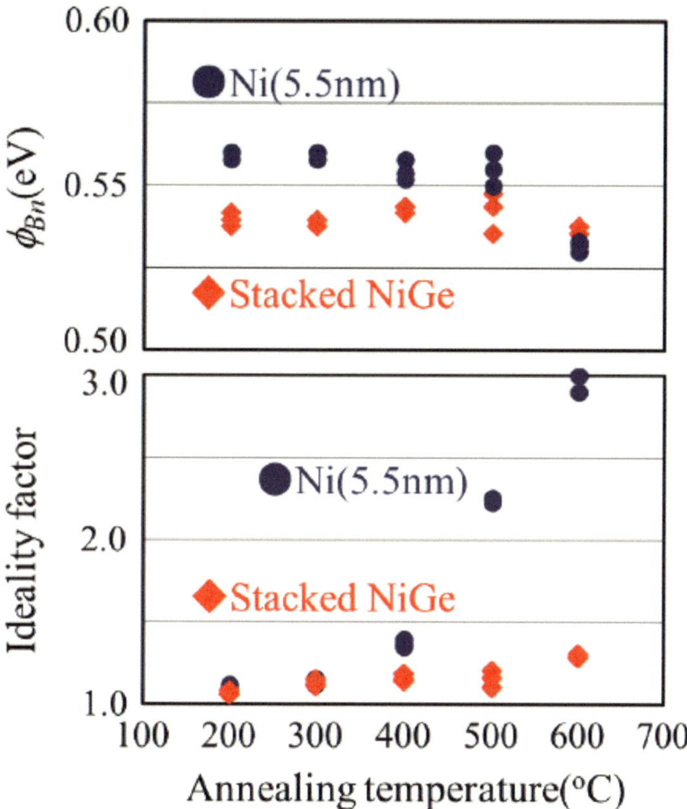

Figure 4. Heights of the potential barrier and the ideality n-factors for the Schottky contacts with stacked NiGe and with a 5.5 nm-thick Ni film, at various annealing temperatures [12].

sample with a 5.5 nm Ni film and no cyclic stacking, the values of the *n*-factor were very large for temperatures above 400°C, owing to thermal instability of NiGe films in this sample. This result is consistent with the sheet resistivity result presented in **Figure 2**. In **Figure 2**, we see that the sheet resistivity for this sample rises rapidly around 400°C.

2.2.2. *Interface dopant incorporation*

A sample with 22 nm of Ni on an *n*-type Ge substrate and another with 30 nm of Ni_3P on an *n*-type Ge were prepared by plasma deposition. **Figure 5** shows a schematic illustration of the two samples.

Current-voltage characteristics were obtained at various annealing temperatures (for 1 min) for the two samples in order to extract the Schottky potential barrier heights, Φ_{Bn} using the thermionic emission model. The results are shown in **Figure 6**.

The ideality *n*-factor for the sample with a 22-nm-thick Ni film was also determined and is presented in **Figure 6**. It can be seen in **Figure 6** that the determination of Φ_{Bn} and the *n*-factor was repeated a number of times at each temperature, showing some experimental scattering error. It is clear that the values of Φ_{Bn} for the Ni/*n*-Ge sample remained almost constant over the whole temperature range of annealing. It is however seen that the values of Φ_{Bn} for the Ni_3P/*n*-Ge sample gradually decreases with increased temperature and the lowest Φ_{Bn} value is achieved at 600°C, after which the value increases. The ideality factor of the Ni/*n*-Ge sample gradually increases but does not go above the value of 1.5.

Since we see from **Figure 6** that the values of Φ_{Bn} for the Ni_3P/*n*-Ge sample are the lowest at 600°C, we now focus on the current-voltage characteristics at this temperature alone for both the Ni/*n*-Ge and Ni_3P/*n*-Ge samples, the results are shown in **Figure 7**.

We see in **Figure 7(a)** that the current density profiles on a semilogarithmic scale for the forward and reverse bias (negative voltage region) directions are symmetric about the zero anode voltage axis for the Ni_3P/*n*-Ge sample, suggesting Ohmic characteristics. On the other hand, for the Ni/*n*-Ge samples, in the reverse bias direction, the current density is very small compared to that in the forward bias direction, showing that the contact is rectifying. In **Figure 7(b)**, the current density profile is presented on a linear scale. For the Ni_3P/*n*-Ge sample, we get a straight line which confirms that this is an ohmic contact. The corresponding result for the Ni/*n*-Ge sample clearly shows that it is not an ohmic contact. It shows Schottky diode

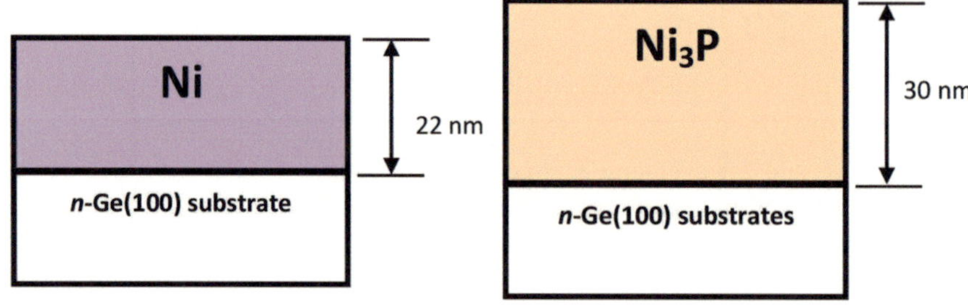

Figure 5. Schematic illustration of a sample with 22 nm of Ni on an *n*-type Ge substrate and another with 30 nm of Ni_3P on *n*-type Ge.

http://dx.doi.org/10.5772/intechopen.79318

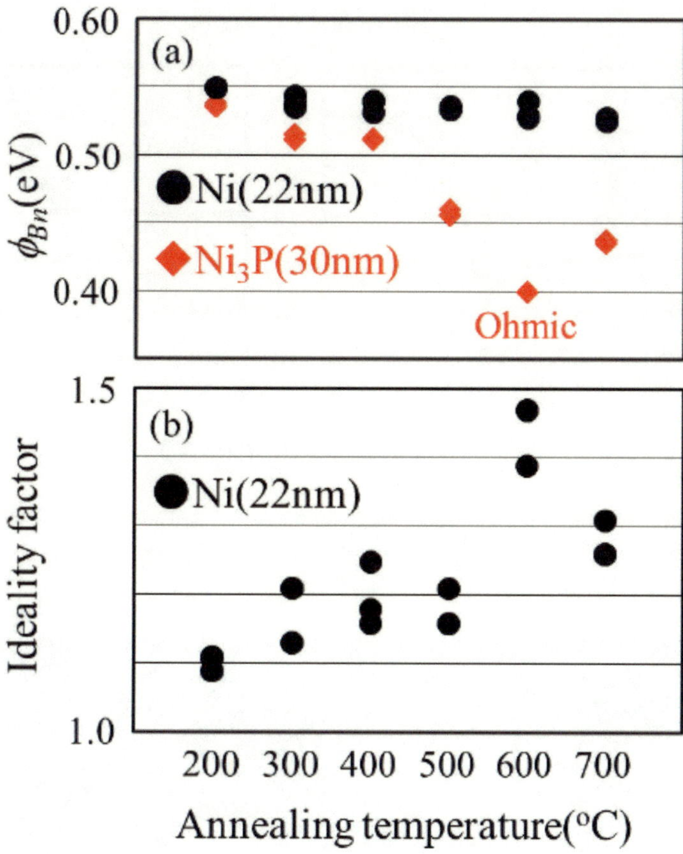

Figure 6. (a) Potential barrier heights at various annealing temperatures (for 1 min) for Ge Schottky contacts with a 30 nm-thick Ni_3P film and with a 22 nm-thick Ni film. (b) The ideality *n*-factor for the sample with a 22 nm-thick Ni film at various annealing temperatures [12].

characteristics as was illustrated schematically in **Figure 11** of the previous chapter. **Figure 8** shows a cross-sectional scanning electron microscope (SEM) image of the Ni_3P (30 nm)/*n*-Ge contact in the as-deposited state, (a) and after annealing at 600°C for 1 min, (b).

The SEM micrograph in **Figure 8(a)** shows a regular thickness of Ni_3P in the as-deposited contact. **Figure 8(b)** shows a much thicker reaction region up to a depth of 77.8 nm below the original interface. We saw in **Figure 6** that ohmic behavior was not achieved for annealing temperatures that were less than 600°C. It appears that at 600°C, the P atoms penetrated enough into the *n*-Ge substrate to form an interface region inside the substrate which is heavily doped by P atoms. This facilitates for the ohmic characteristics observed at 600°C.

2.2.3. Cyclically stacked NiGe contacts with interface dopant incorporation

A thin 0.68 nm film of Ni_3P was plasma deposited on an *n*-type Ge substrate after which sets of Ni/Ge (0.5 nm/1.3 nm) layers were cyclically stacked seven times. **Figure 9** shows a schematic illustration of the sample.

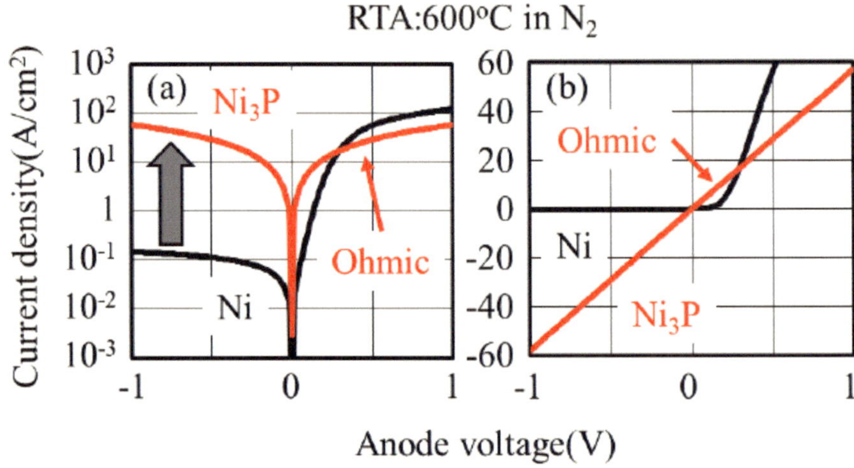

Figure 7. (a) Current density profile on a semilogarithmic scale for the forward and reverse bias directions for Ni/*n*-Ge and Ni_3P/*n*-Ge samples. (b) Current density profile of the same samples on a linear scale [12].

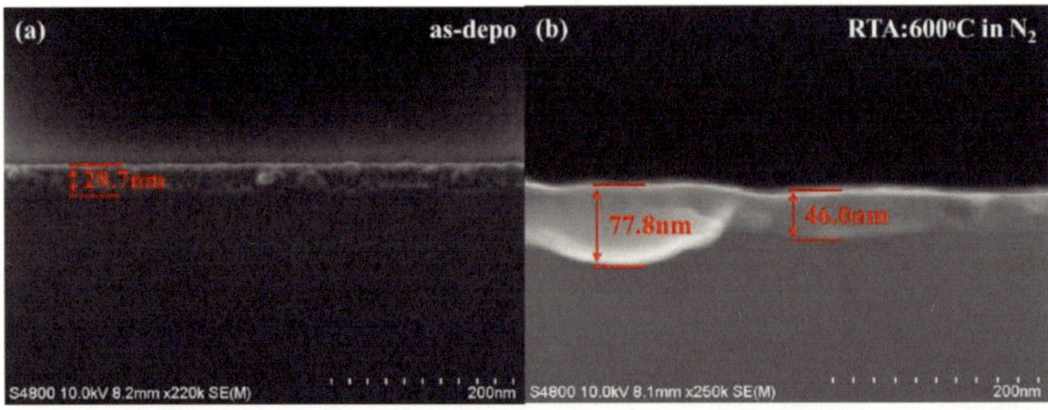

Figure 8. (a) A cross-sectional scanning electron micrograph of an as-deposited Ni_3P/*n*-Ge contact. (b) A micrograph similar to the one presented in (a) but after annealing at 600°C [12].

Current-voltage characteristics were obtained at various annealing temperatures for 1 min. The potential barrier heights, Φ_{Bn} were extracted using the thermionic emission model, as explained earlier. The results are shown in **Figure 10**, and results for cyclically stacked NiGe without a thin 0.68 nm film of Ni_3P are also included in **Figure 10** for comparison.

It is seen in **Figure 10** that the Ni_3P film reduces the barrier height by about 0.51 at 500°C. We now focus on the current-voltage characteristics at the temperatures of 400 and 500°C for both the cyclically stacked samples with and without an Ni_3P film. The results are shown in **Figure 11**.

The sample without an Ni_3P film is used as a control to compare with the one with an Ni_3P film, the results from this sample are therefore labeled as "control" in **Figure 11**. It can be seen

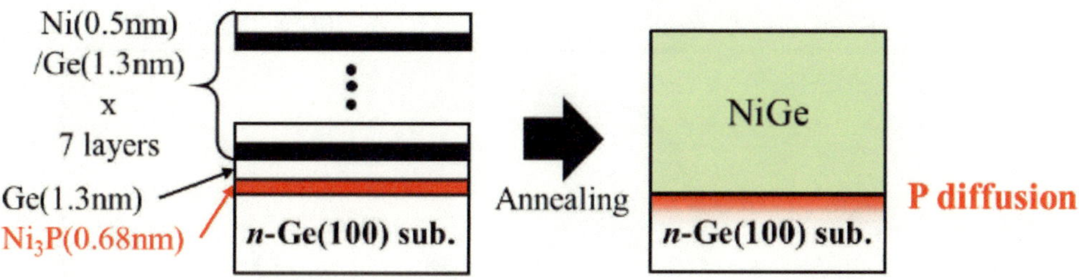

Figure 9. Schematic illustration of sets of Ni/Ge (0.5 nm/1.3 nm) layers cyclically stacked seven times over a thin 0.68 nm film of Ni_3P on an *n*-Ge(100) substrate [12].

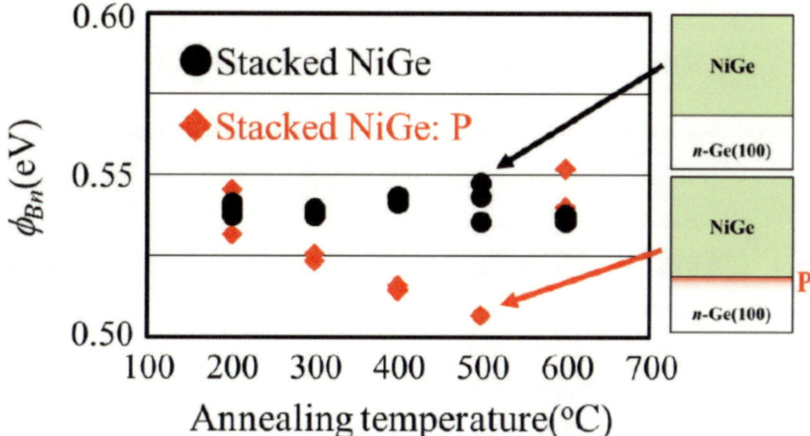

Figure 10. Potential barrier heights extracted at various temperatures for cyclically stacked NiGe with and without a thin 0.68 nm film of Ni_3P [12].

that both contacts are rectifying. Despite the reduction in the barrier height seen in **Figure 10**, the incorporation of a Ni_3P film does not result in an ohmic contact in this case.

2.2.4. Interface insertion of a silicon film

A silicon film with a varying thickness, x, is deposited on an *n*-type Ge substrate after which sets of Ni/Ge (0.5 nm/1.3 nm) layers were cyclically stacked seven times. Samples were prepared in this way for four different values of the Si film thickness, which are: $x = 0.1$, 0.3, 0.6 and 1.0 nm. Another set of similar samples were prepared with the only difference being the introduction of a 0.68-nm-thick Ni_3P film between the Si film and the sets of Ni/Ge (0.5 nm/1.3 nm) layers. **Figure 12** shows a schematic illustration of the two types of sample configuration.

The current-voltage characteristics for samples without P incorporation (w/op) and with P incorporation (w/p), annealed at a temperature of 400°C for 1 min, are shown in **Figure 13** for the different values of Si thickness, x.

It can be seen in **Figure 13** that ohmic characteristics are observed for the contact with P incorporation (w/p) and a silicon film thickness of 0.1 nm. Current-voltage characteristics

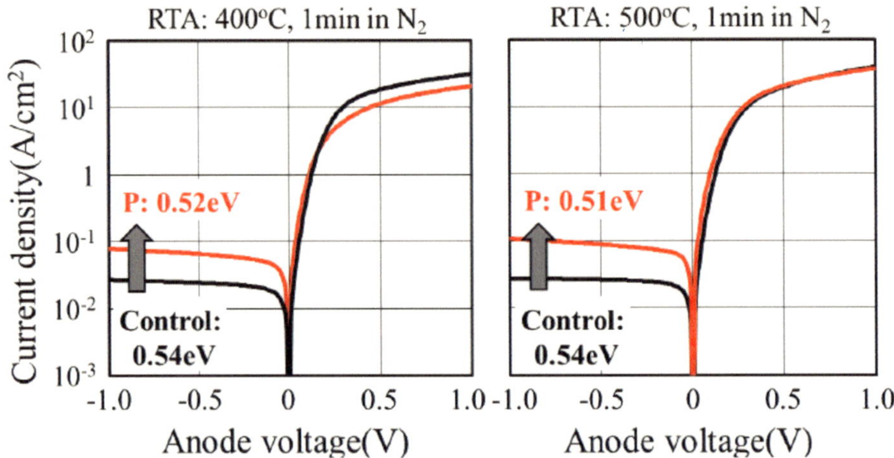

Figure 11. Current-voltage characteristics at the temperatures of 400 and 500°C for both the cyclically stacked samples with and without an Ni_3P film [12].

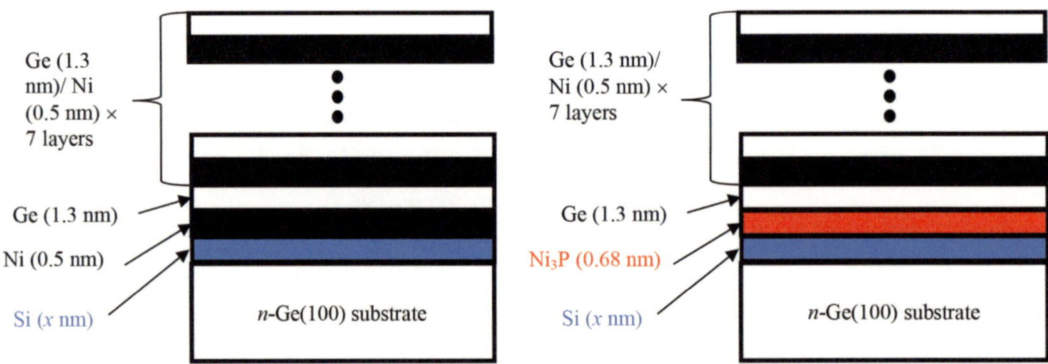

Figure 12. Schematic illustration of two types of cyclically stacked sample configurations with the insertion of a silicon film.

were obtained for annealing temperatures between 200 and 600°C. The barrier heights were determined at these temperatures of annealing and are presented in **Figure 14**.

We see in **Figure 14** that a silicon thickness of 0.1 nm gives the lowest barrier height for both types of samples. However, ohmic characteristics are only observed when P is incorporated and at an annealing temperature of 400°C, as seen in **Figure 13**. An annealing temperature of 300°C with $x = 0.1$ nm gives characteristics that are nearly ohmic as indicated in **Figure 14**.

The four materials, Si, NiGe, *n*-type Ge, and P doped *n*-type Ge have different work functions. These four materials therefore have individual energy band structures with the respective Fermi levels being at different positions relative to the vacuum level. After a contact between any of these materials is made, the Fermi level becomes constant throughout the system at equilibrium, and the energy bands, which should have continuous characteristics, therefore bend.

http://dx.doi.org/10.5772/intechopen.79318

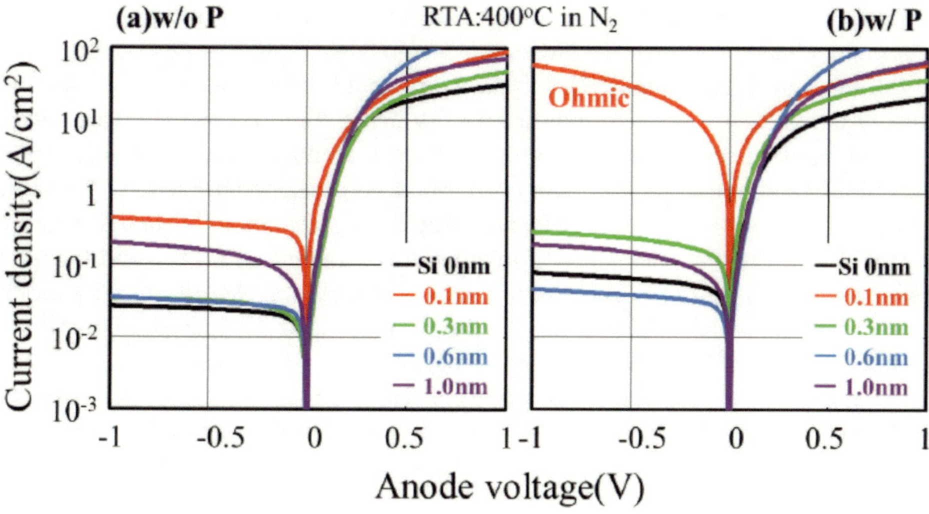

Figure 13. Current-voltage characteristics for samples without P incorporation (a) and with P incorporation (b), annealed at a temperature of 400°C for 1 min and different values of Si thickness, x [12].

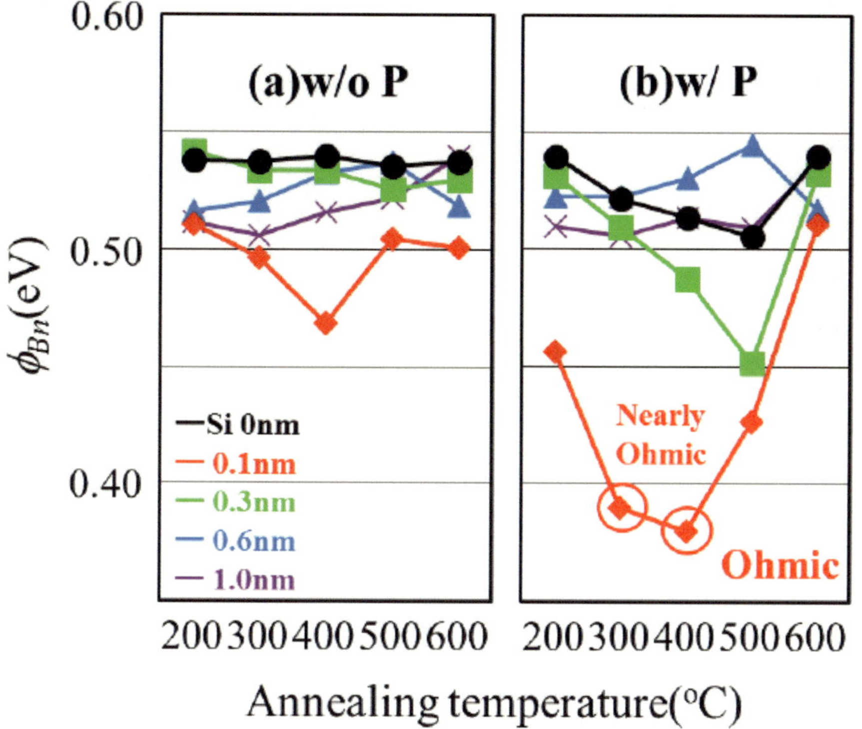

Figure 14. (a) Barrier heights at various temperatures of annealing and varying thickness of the Si film inserted, without the incorporation of P atoms. (b) A plot of results similar to the ones in (a) but with the incorporation of P atoms [12].

Figure 15(i) shows this energy band bending at the NiGe/*n*-Ge interface. **Figure 15(ii)** and **(iii)** show the energy band bending at the NiGe:P/*n*-Ge and NiGe:P/Si/*n*-Ge interfaces respectively. We see the successive reduction in the Schottky potential barrier heights from 0.54 to 0.51 eV in **Figure 15(i)** and **(ii)** respectively and then 0.42 eV in the ohmic contact of **Figure 15(iii)**. The energy band bending due to the doping by diffused P atoms is shown in **Figure 15(ii)** and **(iii)**. **Figure 15(i)** shows that the Fermi level is pinned. As explained in Section 1.1.5 of the previous chapter, this Fermi-level pinning is caused by dipole formation due to the large amount of interface states. The diameter of a silicon atom is approximately 0.1 nm and the thickness of the silicon layer in **Figure 15(iii)** is also 0.1 nm, meaning that it was essentially a monolayer deposition of silicon which released the Fermi-level pinning and achieved an ohmic contact. The large modulation of the Schottky potential barrier height due to the insertion of a 0.1-nm-thick Si monolayer is explainable by considering that the Si atoms caused the formation of chemical bonds between NiGe, Si, and *n*-Ge, thereby reducing the number of dangling bonds (valence mending) that are responsible for the dipole formation which is explained in Section 1.1.5 of the previous chapter.

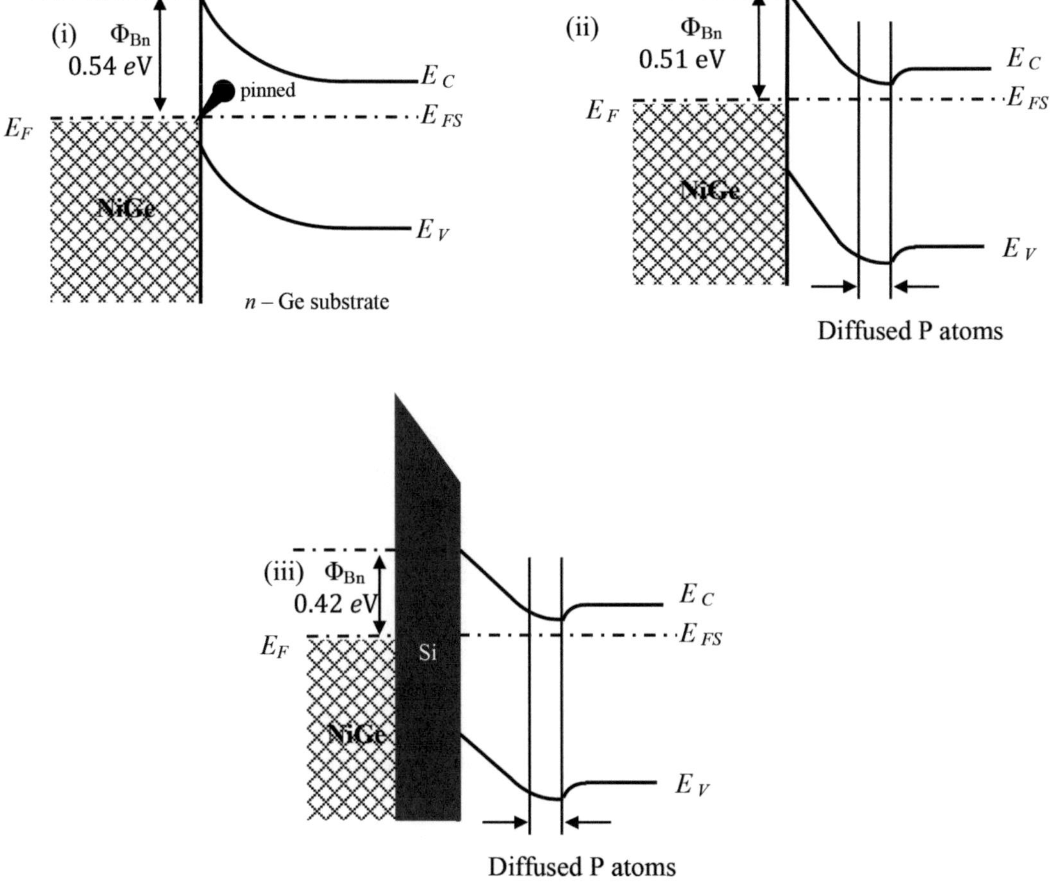

Figure 15. Schematic energy band diagrams of: (i) an NiGe/*n*-Ge interface; (ii) an NiGe/*n*-Ge interface with the incorporation of P atoms; (iii) an NiGe/*n*-Ge interface with the incorporation of P atoms and the insertion of a silicon layer.

2.3. PdGe contacts

Chawanda et al. [13] used *n*-type Ge(111) doped with antimony (Sb) at a density of 2.5×10^{15} cm^{-3}. Pd was deposited onto the substrates by vacuum resistive evaporation as explained in Section 2.1.1 of the previous chapter. This was done through a mechanical mask which had circular windows of diameter, 0.6 ± 0.05 mm. In this way, 24 circular contacts were prepared in a single evaporation. The thickness of each deposited layer of Pd was 30 nm. Room temperature forward and reverse bias current-voltage characteristics were obtained for five of the Pd/*n*-Ge contacts, which acted as Schottky barrier diodes. The results are shown in **Figure 16**.

Rectifying characteristics are seen for all samples in **Figure 16**. The height of the Schottky potential barrier, Φ_{Bn} and the ideality *n*-factor were extracted from the forward bias I-V characteristics of the Schottky diodes at room temperature using the thermionic emission model, as explained in Section 1.1.4 of the previous chapter. This was done for several samples and a histogram was produced to show the statistical distribution of the effective potential barrier heights from the forward bias I-V characteristics, and this is presented in **Figure 17**.

The effective potential barrier heights obtained from the I-V characteristics varied from 0.492 to 0.550 eV. A Gaussian distribution function was used to obtain fits to the histogram. The statistical analysis yielded a mean Schottky potential barrier height value of 0.529 eV with a standard deviation of 0.019 eV.

A histogram was also produced for the values of the ideality factors determined from the I-V characteristics. **Figure 18** shows the statistical distribution of ideality factors from the forward bias I-V characteristics.

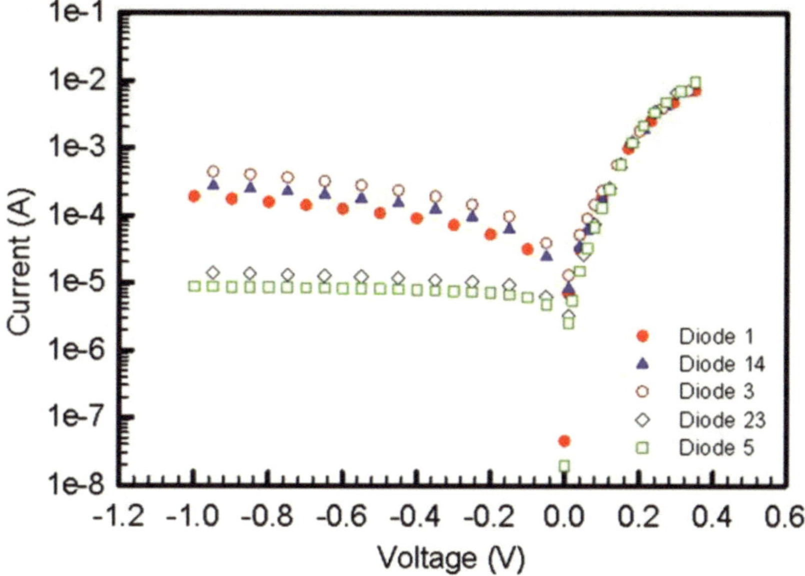

Figure 16. Room temperature forward and reverse bias current-voltage characteristics obtained for five of the Pd/*n*-Ge contacts [13].

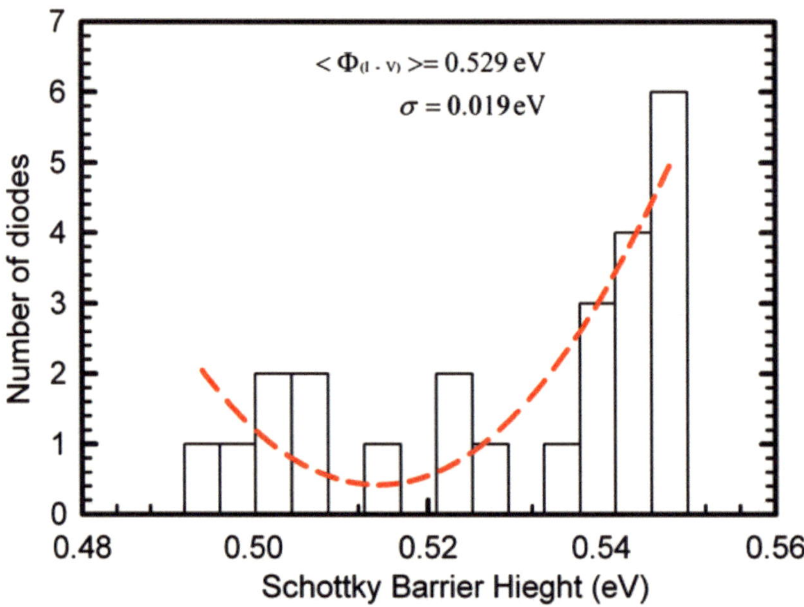

Figure 17. Room temperature Schottky potential barrier height distribution derived from forward bias I-V characteristics [13].

The ideality factor ranged from 1.140 to 1.950. A Gaussian distribution function was used to obtain a fit to the histogram. The statistical analysis of the ideality factor yielded an average value of 1.414 with a standard deviation of 0.270.

It is seen that the experimental effective potential barrier heights and ideality factors differ from contact to contact even though they were identically prepared in a single evaporation and on the same substrate. A plot of the effective potential barrier heights as a function of the respective ideality factors is shown in **Figure 19**.

The experimental effective potential barrier height decreases as the ideality factor increases. We see a linear relationship and the straight line drawn in the figure is the least-squares fit to the experimental data.

Five Pd/*n*-Ge contacts were experimentally examined at room temperature to obtain the reverse bias C^{-2}-V characteristics. The results are shown in **Figure 20**.

We see in **Figure 20** that each contact gives a straight line in the C^{-2}-V graphs. The value of the capacitance-voltage derived potential barrier height, $\Phi_{B(C-V)}$ can be obtained from **Figure 20** using,

$$\Phi_{B(C-V)} = V_D + E_F - \Delta\Phi_B \tag{1}$$

where E_F is the energy difference between the bulk Fermi level of Ge and the conduction band edge, V_D is the diffusion potential, and $\Delta\Phi_B$ is the image-force barrier lowering given by Eq. (16) in Section 1.1.4 of the previous chapter, where the maximum electric field E_m is calculated using Eq. (18) of the previous chapter.

The reverse bias C^{-2}-V characteristics were obtained for several samples and a histogram was produced to show the statistical distribution of the capacitance-voltage-derived potential barrier heights, and this is presented in **Figure 21**.

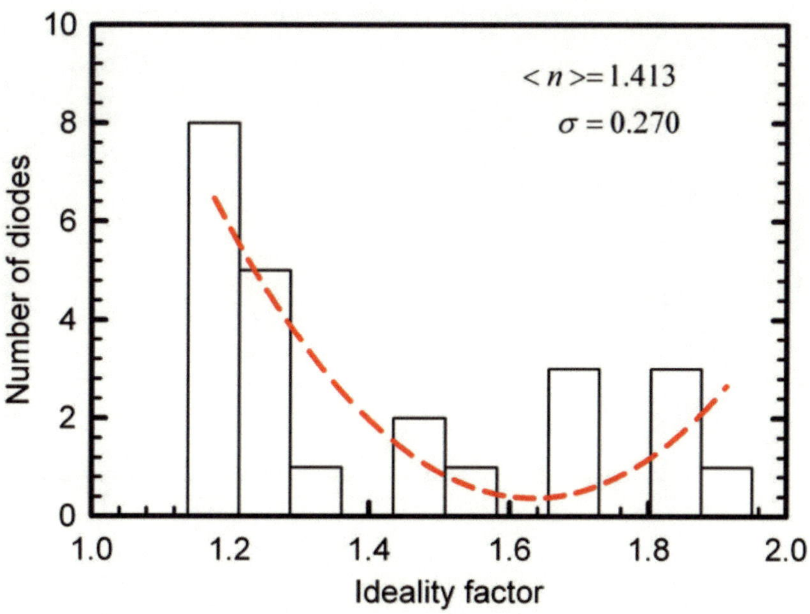

Figure 18. Statistical distribution of ideality factors from the forward bias I-V characteristics [13].

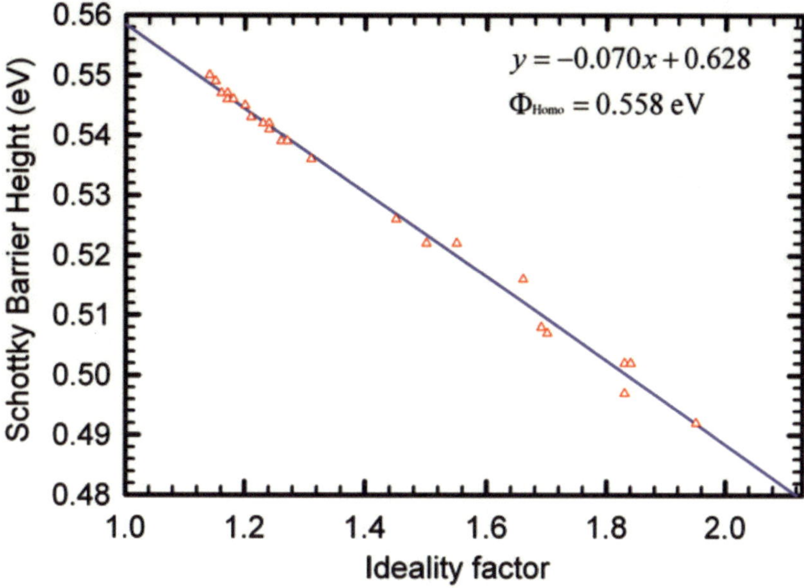

Figure 19. A linear plot of the Schottky potential barrier heights against the ideality factors [13].

The capacitance-voltage potential barrier heights for the Pd/*n*-Ge (111) Schottky structures varied from 0.427 to 0.509 eV. The statistical analysis yields a mean barrier height of 0.463 eV with a standard deviation of 0.023 eV. The difference between the mean Schottky potential barrier height obtained using the C^{-2}–V characteristics and that from the I-V characteristics

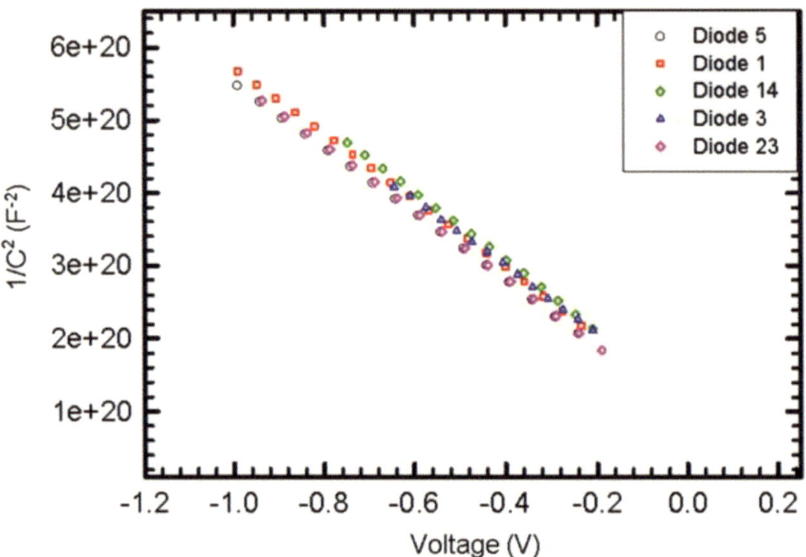

Figure 20. Schottky reverse bias C^{-2}-V characteristics for five Pd/n-Ge samples [13].

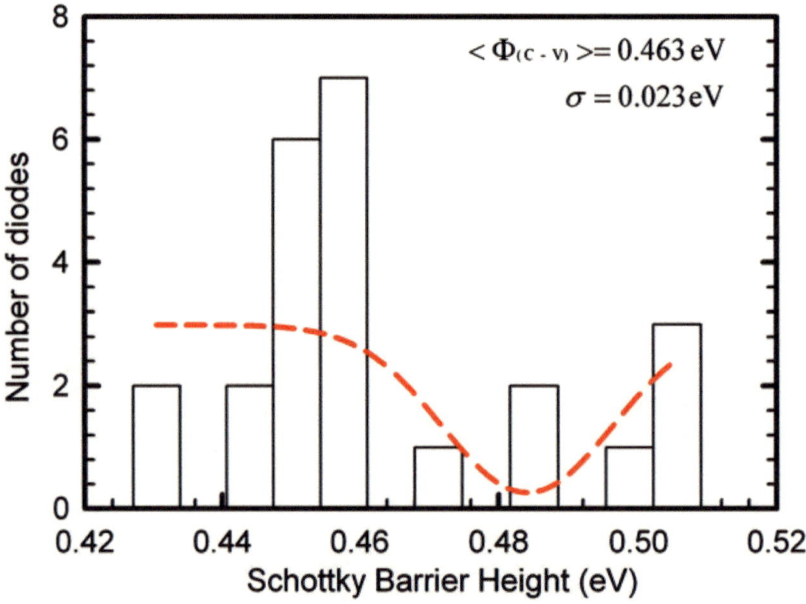

Figure 21. Schottky potential barrier height distribution derived from reverse bias C^{-2}-V characteristics [13].

is 0.066 eV. Potential barrier heights obtained from the I-V and C^{-2}-V characteristics are not always the same [14] because of the different nature of the two measurement techniques.

2.3.1. *Interface dopant implantation*

Descoins et al. [15] used Ge (001) substrates to form two types of Pd/Ge contacts. In the first type of samples the surface of the substrates were implanted with Se atoms at an energy of 130 keV as explained in section 2.2 of the previous chapter. The samples were then vacuum annealed at a pressure of 4 × 10^{-5} Torr using a rapid thermal annealing (RTA) setup at 700°C for 30 min. This annealing was done to activate some diffusion of the Se dopant atoms further into the semiconductor surface, before metallization with Pd. A Pd film with a thickness of 20 nm was then deposited at room temperature onto the surface of the sample using magnetron sputtering at a base pressure of 10^{-8} Torr. The second type of samples was prepared in exactly the same way as the first type but with no Se implantation and activation. All the samples were then vacuum annealed at a pressure of 10^{-6} Torr to induce solid state reactions, resulting in the formation of the PdGe phase. X-ray diffraction (XRD) measurements were made in-situ. The heating ramp rate was 5°C per min steps and these steps were separated by 5 min-long XRD measurements at a constant temperature.

For the samples which were implanted with Se, the distribution of Se atoms in the surface region was determined at three stages of the sample preparation using secondary ion mass spectrometry (SIMS). The distribution was first obtained immediately after the Se implantation (as implanted). It was also obtained after the rapid thermal annealing at 700°C for 30 min, which was done to activate some diffusion of the Se dopant atoms. The third SIMS determination of the Se distribution was carried out after the annealing ramp which resulted in the formation and growth of the PdGe phase. All secondary ion mass spectrometry results are presented in **Figure 22**. The Se SIMS profile measured immediately after the Se implantation is represented by open triangles. The profile after the activation annealing was performed at 700°C for 30 min and is represented by the open squares.

The Se SIMS profile measured after the annealing ramp to form PdGe is represented by open circles in **Figure 22**. If we compare this profile to the one obtained after the activation annealing was performed at 700°C for 30 min (open squares), we see that Se atoms did not diffuse any further into the depth of the substrate during the annealing ramp. The Se profile immediately after the implantation corresponds to a Gaussian distribution with a maximum concentration

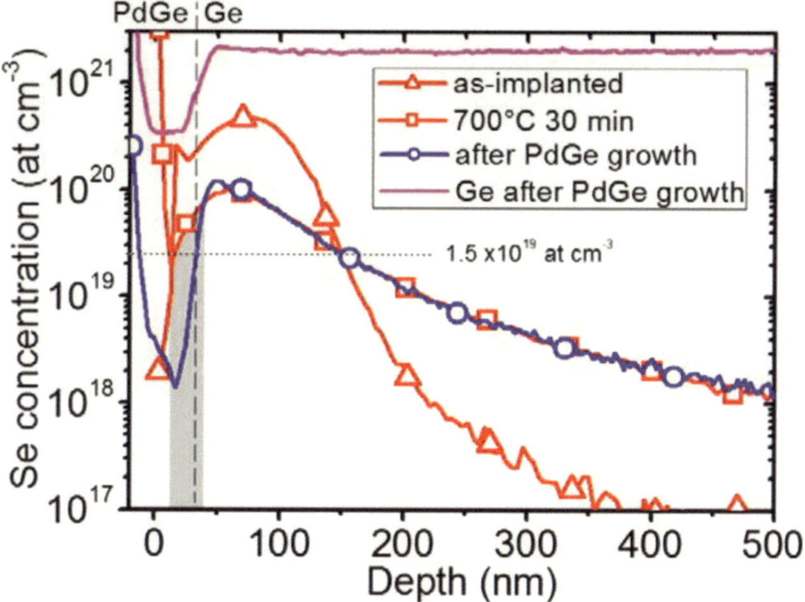

Figure 22. Secondary ion mass spectrometry results [15].

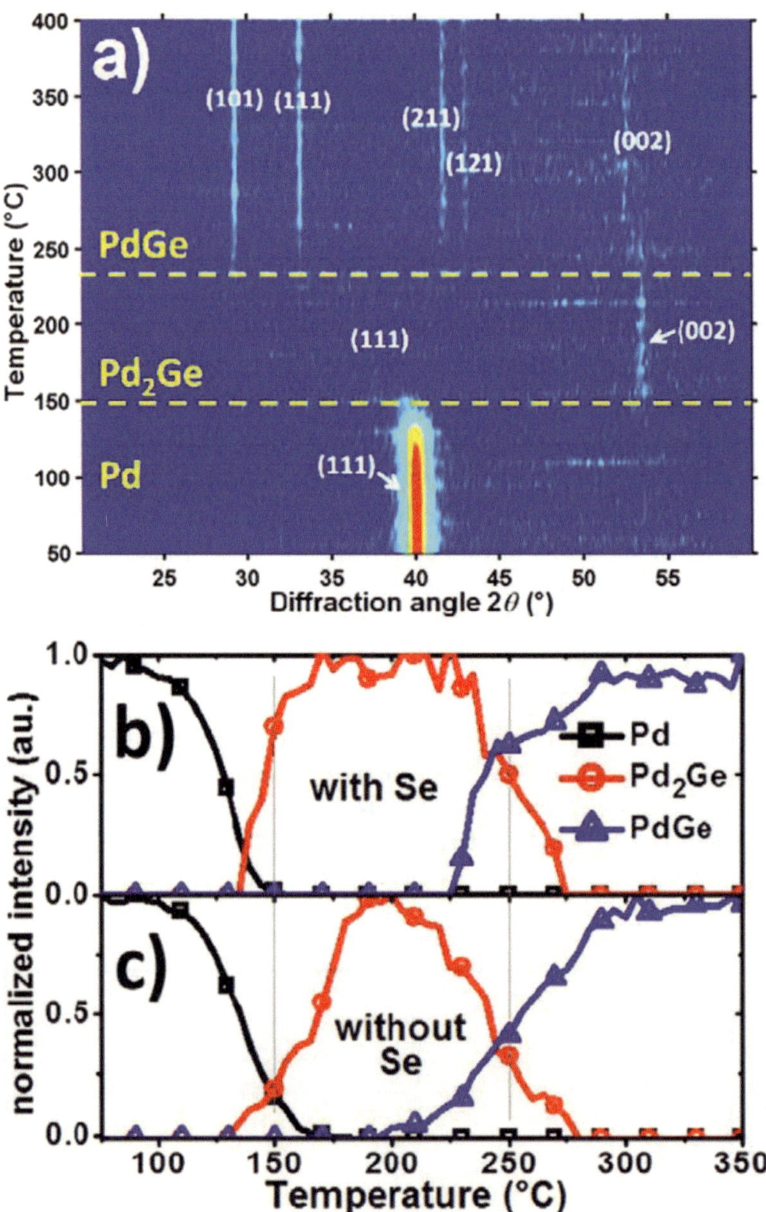

Figure 23. (a) In-situ X-ray diffractogram obtained from an Se-doped sample. (b) Pd(111), Pd_2Ge(002) and PdGe(101) integrated and normalized XRD peak data extracted in-situ during the annealing of a sample with Se doping. (c) Integrated and normalized XRD peak data for a sample without Se doping [15].

of about 5×10^{20} atoms cm^{-3}, which is located at around 60 nm below the surface of the sample. As a result of the activation annealing, the Se atoms diffused further into the substrate decreasing the maximum Se concentration at a depth of 60 nm from 5×10^{20} to about 1×10^{20} at cm^{-3}.

Figure 23(a) shows the in-situ X-ray diffractogram obtained from an Se-doped sample. This diffractogram evolved during the in-situ XRD annealing process. Initially only a single Pd(111) diffraction peak is detected at a diffraction angle of $2\theta \approx 40°$. Upon annealing, the $Pd_2Ge(111)$ and $Pd_2Ge(002)$ peaks appeared at $2\theta \approx 37.5$ and 53.7°, respectively. The intensity of the Pd(111) peak decreased during further annealing and that of the $Pd_2Ge(111)$ and Pd_2Ge (002) peaks increased until the Pd(111) peak disappeared after which five new peaks corresponding to the PdGe(101), (111), (211), (121), and (002) planes appeared. The Pd_2Ge then starts to get consumed, giving way to PdGe growth. This evolution is displayed in **Figure 23(b)** for the sample with Se doping and in **Figure 23(c)** for the sample without Se doping. To get the results in **Figure 23(b)** and **(c)**, the XRD peak intensities corresponding to various phases were integrated and normalized. The normalized integrated intensities were then plotted against the temperatures of the ramp annealing. We see from **Figure 23(b)** and **(c)** that at the end of the experiment we have a layer of PdGe in contact with an Se-doped Ge substrate and another in contact with an Se-free Ge substrate. Sheet resistivity measurements were carried out on both, the samples with Se interface doping and those without Se doping. The resistivity of the PdGe film grown on the Se-free Ge substrate was found to be, $\rho_{sh} = 13 \pm 1$ μΩ cm and that on the Ge substrate with interface Se doping was, $\rho_{sh} = 6 \pm 0.8$ μΩ cm. This means that interface Se doping reduces the sheet resistivity by half which should result in a nearly ohmic contact because the sheet resistivity is closely related to the contact resistivity.

3. Summary and conclusion

Some of the novel interface control processes developed for the fabrication of NiGe and PdGe Schottky and ohmic contacts on *n*-type germanium have been reviewed.

NiGe grown using the cyclic stacking of Ni/Ge films on an *n*-Ge substrate showed a stable sheet resistivity in the annealing temperature range from 275 to around 500°C. This temperature range was much wider than the corresponding stable-sheet-resistance annealing temperature range obtained from NiGe grown under similar conditions but without the cyclic stacking. The Schottky potential barrier heights for the contacts with cyclically stacked NiGe exhibited stable values which were less than 0.54 eV, even after annealing at temperatures of up to 600°C. The ideality factors of these contacts were less than 1.2, even after annealing at temperatures of up to 500°C. NiGe contacts with the interface incorporation of phosphorus atoms and the insertion of a silicon film at the interface were explained. Ohmic characteristics have been observed for contacts with substantial P interface incorporation and those with minimal P interface incorporation coupled with the insertion of a 0.1 nm-thick Si film, which is essentially a monoatomic Si layer.

A linear relationship was observed between the potential barrier heights and corresponding ideality factors for Schottky contacts of Pd grown on Sb-doped Ge(111) with a doping density of about 2.5×10^{15} cm^{-3}. Current-voltage and capacitance-voltage characteristics were obtained at room temperature. The effective potential barrier heights obtained from these I-V characteristics varied from 0.492 to 0.550 eV, while the ideality factor varied from 1.140 to 1.950. The barrier heights obtained from the reverse bias capacitance-voltage (C^{-2}-V) varied from 0.427 to 0.509 eV. A Gaussian distribution function was fitted over the experimental potential barrier height distributions, resulting in average values of 0.529 and 0.463 eV from I-V and C^{-2}-V characteristics, respectively.

The sheet resistivity of PdGe grown by Pd reactive diffusion on Ge substrates which had their surfaces implanted with Se atoms was two times lower than that for samples grown under the same conditions but without Se implantation. This result suggests that Se atoms at the Pd/*n*-Ge interface may be used to produce efficient PdGe contacts. The presence of the Se atoms does not modify either the phase-formation sequence or the phase growth kinetics during the Pd reactive diffusion with the Ge substrate, as seen in **Figure 23**.

The three interface control processes analyzed, namely the interface incorporation of P atoms, the thin film insertion of Si at the interface, and the implantation of Se atoms into the surface of the semiconductor substrate, have been demonstrated to be effective, and are therefore recommended for the improvement of Ni and Pd contacts in the next generation of *n*-type germanium-based nanoelectronic devices.

Acknowledgements

The author would like to thank the Copperbelt University for the use of the institution's facilities.

Author details

Adrian Habanyama

Address all correspondence to: adrian.habanyama@cbu.ac.zm

Department of Physics, Copperbelt University, Kitwe, Zambia

References

[1] Brunco DP, De Jaeger B, Eneman G, Mitard J, Hellings G, Satta A, Terzieva V, Souriau L, Leys FE, Pourtois G, Houssa M, Winderickx G, Vrancken E, Sioncke S, Opsomer K, Nicholas G, Caymax M, Stesmans A, Van Steenbergen J, Mertens PW, Meuris M, Heyns MM. Germanium MOSFET Devices: Advances in Materials Understanding, Process Development, and Electrical Performance. Journal of the Electrochemical Society. 2008;**155**(7):H552-H561

[2] Gaudet S, Detavernier C, Kellock AJ, Desjardins P, Lavoie C. Thin film reaction of transition metals with germanium. Journal of Vacuum Science and Technology. 2006;**A24**:474

[3] Jin LJ, Pey KL, Choi WK, Fitzgerald EA, Antoniadis DA, Pitera AJ, Lee ML, Chi DZ, Tung CH. The interfacial reaction of Ni with (111)Ge, (100)Si0.75Ge 0.25 and (100)Si at 400 °C. Thin Solid Films. 2004;**462-463**:151-155

[4] Nemouchi F, Mangelinck D, Bergman C, Clugnet G, Gas P. Simultaneous growth of Ni_5Ge_3 and NiGe by reaction of Ni film with Ge. Applied Physics Letters. 2006;**89**:131920

[5] Patterson JK, Park BJ, Ritley K, Xiao HZ, Allen LH, Rockett A. Kinetics of Ni/a-Ge bilayer reactions. Thin Solid Films. 1994;**253**(1-2):456-461

[6] Gaudet S, Detavernier C, Lavoie C, Desjardins P. Reaction of thin Ni films with Ge: Phase formation and texture. Journal of Applied Physics. 2006;**100**:034306

[7] Majni C, Ottaviani G, Zani A. On the growth kinetics and structure of Pd_2Ge and PdGe. Journal of Non-Crystalline Solids. 1978;**29**(3):301-309

[8] Knaepen W. Characterization of solid state reactions and crystallization in thin films using in situ X-ray diffraction [PhD thesis]. University of Ghent; 2010

[9] Hökelek E, Robinson GY. A comparison of Pd Schottky contacts on InP, GaAs and Si. Solid State Electronics. 1981;**24**(2):99-103

[10] Comrie CM, Pondo KJ, van der Walt C, Smeets D, Demeulemeester J, Habanyama A, Knaepen W, Detavanier C, Vantomme A. Determination of the dominant diffusing species during nickel and palladium germanide formation. Thin Solid Films. 2012;**526**:261-268

[11] Chilukusha D, Pineda-Vargas CA, Nemutudi R, Habanyama A, Comrie CM. Microprobe PIXE study of Ni-Ge interactions in lateral diffusion couples. Nuclear Instruments and Methods in Physics Research B. 2015;**363**:161-166

[12] Motoki M. Interface control process for metal/germanium Schottky contact [MSc thesis]. Japan: Tokyo Institute of Technology; 2015

[13] Chawanda A, Roro KT, Auret FD, Mtangi W, Nyamhere C, Nel J, Leach L. Determination of the laterally homogeneous barrier height of palladium Schottkybarrier diodes on n-Ge(111). Materials Science in Semiconductor Processing. 2010;**13**:371-375

[14] Güler G, Karatas S, Güllül Ö, Bakkaloglu ÖF. The analysis of lateral distribution of barrier height in identically prepared Co/n-Si Schottky diodes. Journal of Alloys and Compounds. 2009;**486**(1-2):343-347

[15] Descoins M, Perrin Toinin J, Zhiou S, Hoummada K, Bertoglio M, Ma R, Chow L, Narducci D, Portavoce A. PdGe contact fabrication on Se-doped Ge. Scripta Materialia. 2017;**139**:104-107